WOMEN AND PLANNING

Planning is currently a male-dominated profession. Women have been central to the planning movement since it began.

This book offers a comprehensive account of women and the planning movement past and present, covering philosophical, practical and policy dimensions. The construction of the sanitized 'story' of modern, scientific, town planning hides past links with eugenics, colonialism, artistic, utopian and religious movements and the occult.

Under the pretence of spatial problem solving, male planners are projecting onto urban society their own patriarchal perceptions of what the built environment ought to be like, 'creating urban realities' with little reference to the 'real' needs of the majority of the population. *Women and Planning* discusses the present gendered nature of town planning and questions planners' assumptions concerning the characteristics of those they are planning 'for' in terms of gender, class, sexuality, ethnicity, personal perspectives and disability.

Drawing on the author's own experience as a planner and one of the planned, this book will be of particular interest to students of town planning, urban studies and women's studies, and professionals in town planning and the wider built environment.

Clara H. Greed is Senior Lecturer in the School of Planning at the University of the West of England. Her many publications on the built environment include *Introducing Town Planning* and *Surveying Sisters*.

WOMEN AND PLANNING

Creating gendered realities

Clara H. Greed

London and New York

First published 1994
by Routledge
11 New Fetter Lane, London EC4P 4EE

Transferred to Digital Printing 2004

Simultaneously published in the USA and Canada
by Routledge
29 West 35th Street, New York, NY 10001

Typeset in Garamond by
Ponting–Green Publishing Services, Chesham, Bucks

British Library Cataloguing in Publication Data
A catalogue record for this book is available from
the British Library

Library of Congress Cataloging in Publication Data
Greed, Clara,
Women and planning : creating gendered realities /
Clara H. Greed.
p.   cm.
Includes bibliographical references and index.
1. Women and city planning   I. Title.
HT166.G687   1994
307.1'2'082–dc20   93–41715

ISBN 0–415–07980–2 (hbk)
ISBN 0–415–07981–0 (pbk)

CONTENTS

Preface and acknowledgments   vii

1 INTRODUCTION: Beliefs and realities   1
   Gendered expectations   1
   Is there still a problem?   3
   Context and inter-relationships   6
   The urban question   8
   Theoretical basis   9
   Belief   11
   Conceptualization   12
   Methodology and tone   13
   Unresolved issues   15

2 THE PLANNERS: Powers and limitations   18
   Blaming the planners   18
   Other actors in the development process   19
   Scope and nature of the planning system   22
   Representation of women   25
   Types of planners   29

3 WOMAN IN THE CITY OF MAN   34
   Diversity and definitions   34
   City form and structure   40
   Making interconnections   49

4 PLANNING: The spirit of the age   51
   What's behind planning?   51
   Other manifestations of planning   55
   Designing urban reality   65

5 REFLECTIONS ON THE HISTORY OF PLANNING   70
   The story of planning   70
   Perceptions of the past   73
   Reverting to type   86

v

CONTENTS

6 THE NINETEETH CENTURY 88
   Industrialization or masculinization 88
   The development of malestream planning 92
   Women's planning heritage 95
   Restrictions or revelations 104

7 PROFESSIONAL POWER OVER PRIVATE SPACE 107
   The public organization of planning 107
   Town planning policy 115
   Outside influences: The new technology 121

8 PRODUCTION AND CONSUMPTION: Post-war planning 125
   Images of post-war reality 125
   Commerce and consumption 127
   Controlling urban space 133

9 URBAN SOCIOLOGICAL PERCEPTIONS OF WOMEN 141
   Dichotomized perceptions 141
   Disillusionment with the gospel of planning 146
   A more realistic view of society? 148

10 WOMEN INTO PLANNING: Ways and means 156
   The opportunity of post-war planning education 156
   Zone zapping: The women and planning movement 162
   Manifestations of planning for women 168

11 PLANNING FOR WOMEN 173
   Changing the agenda 173
   Planning differently 174
   Prioritizing safety 177
   Dezoning 178
   Planning law: Putting policy into practice 184
   A gender city? 186
   Conclusion 189

Appendix I: Women in planning 194
Table 1: Membership of the professional bodies as at December 1993 ... 194
Table 2: Women members of the Royal Town Planning Institute 195
Table 3: Gender divisions in planning education in Britain 195
Table 4: Women's employment in local government planning
   departments 196
Table 5: RTPI membership as at December 1993 197
Appendix II: Key texts on women and built environment and planning .. 198
Appendix III: Women and planning: policy proposals and initiatives ... 202

   Bibliography 205
   Name index 235
   Subject and place index 240

PREFACE AND ACKNOWLEDGMENTS

The purpose of this work is not to provide a prescriptive textbook on how to plan for women (for that see Greed, 1993a, chapter 12), but to reflect on the gendered nature of the beliefs planners have espoused in creating urban realities.

My thanks to Professor Alison Ravetz (Leeds), Madge Dresser (University of the West of England, Bristol) and Philippa Levine (University of Southern California) whose comments led me to see the emphasis on public/private dichotomies latent within my work which has developed into a key theme; and to the women in planning group of the Scottish Branch of the Royal Town Planning Institute (RTPI) for commenting on the manuscript.

My thanks to Deborah Barnes, Judith Taylor, Karla Ni Bhride, and the London Boroughs' women and planning group for reference to their work; and to Beverley Taylor, Hazel McKay, Gordon Cherry, Janet Brand, Hilary Howatt, June Jackson, Sule Takmaz, Sue Buckingham-Hatfield, Jane Foulsham, Marjorie Bulos, Patricia Roberts, Maureen Farrish, Catherine Claverton, Patsy Healey and Diana Lamplugh for material. Information from the following RTPI branches was valuable in building up the wider picture: Hilary Howatt and the Scottish Branch; Maeve Barrett of the Irish Planning Institute women and planning group, Dublin; Ruth Cadbury and colleagues in the London women and planning group; Dory Reeves, Christine Booth and Sunethra Mendis of the Yorkshire Branch; Colette Blackett, South East Branch; Margot Duncalfe, Southampton; South Wales colleagues; South West Branch; and in Bristol Angela Crofts and the Bristol Women's Architects Group and to 'Women in Property' South West.

There are many others whose names appear in the bibliography or the name index, plus the course leaders, local authorities, and individual planners – female and male – who responded to my requests, not forgetting Michael Napier and John Philips at the RTPI for policy statements and, statistics, and at the Faculty of the Built Environment, University of the West of England (formerly Bristol Polytechnic), all colleagues and students who have inspired me.

Autumn 1993

I

INTRODUCTION

Beliefs and realities

GENDERED EXPECTATIONS

The Royal Town Planning Institute's 75th Anniversary Brochure (RTPI, 1989a: 11) shows a photograph of the Institute's Presidential Badge worn on the chain of office. Designed in 1924, by a woman, Mrs Winny Austin, (Cherry, 1974: 114) the brochure states the badge 'shows a female "genius of the city" holding the shields of architecture and engineering'. Likewise, the sculpting of a portrait bust of the grandfather of British town planning, Ebenezer Howard, was entrusted to a Miss Ivy Young (Howard, 1898 [1960]: plate opposite p.49). On buying Howard's book in 1967, having just entered a town planning course, after spending one year at a predominantly female art college studying three dimensional design, I identified with the woman in the photograph, imagining her at work creating Ebenezer Howard. But, when I embarked on a town planning degree, at an overwhelmingly male white-heat-of-technology 1960s university[1], I found that the apparent prominence of women in the iconography of the profession was misleading. The presidential badge has been worn by only one woman in the history of the profession, Sylvia Law (1974–5). Hazel McKay became the second woman president, in 1994. In a profession so dominated by the influence of the forefathers of town planning such as Ebenezer Howard and Le Corbusier, no women have gained equal recognition in urban design. Indeed, many feminists argue it is impossible for there to be such a thing as a female genius within patriarchal society (Battersby, 1989).

As a child in the Fifties, I imagined town planning was to do with better housing, sunshine and gardens. I used to delight in cutting out the Rinso Model Village – houses and shops printed on the back of the cardboard washing powder packets produced by Lever Brothers, the soap manu-facturers.[2] Possibly the aim of this campaign was to socialize little girls into becoming good housekeepers. If so, I misunderstood, because the models increased my interest in buildings and town planning, not in doing the washing. The Lever family were pioneers in town planning, building the model industrial town of Port Sunlight, and endowing the first town planning

1

chair, of Civic Design at Liverpool University in 1909 (Ashworth, 1968: 193). I was fascinated by the idea of lawned front gardens, which were not a common feature in my part of South London. I determined that one day I would live in the suburbs, or possibly Australia, where, according to the publicity material encouraging people to emigrate for only £10, houses had front gardens and the sun always shone (as described by Teather, 1991a: 476). Later I found individual gardens were seen by some town planning purists as elements contributing to the evils of sprawl (Oliver et al., 1981; Hall, 1977a), and suburban houses were perceived as 'ugly little man traps' as J.B. Priestley put it in 1934 (Casson, 1978). Gurus of the modern architectural movement favoured 'returning' to higher densities (many had never left).

Like many girls of the post-war baby boom, I had high ambitions and a desire to solve every problem I met. I would daydream in the mid-1950s about how I would solve the traffic problems of London and the South East so that fewer cats got run over, and fewer people were knocked off their bicycles (in retrospect, what traffic problems?). As we sped along in the family car on our weekly escape to Shoreham-by-Sea, hated by elitist planners as a 'plotlands bungalow town' (King, 1984: 110), I remember thinking if the government built more roads there would be no land left in between. Ten years later, on entering a town planning course at university, I was surprised to find that the objective of transportation planning was not to reduce the numbers of injured cats and mangled bicycles, or time spent waiting at bus stops, but, it seemed, to increase the dominance of the motor car. Meanwhile, in my early teens we had moved down in the world to an inner city area (immortalized in Morris, 1958) where we were constantly subjected to the machinations of the planning system, to threats of road widening, the destruction of small businesses, slum clearance, and to fears of being compulsorily purchased and being rehoused in a grotty high rise block. One day Mum was out shopping and happened upon a young man with a clipboard, who told her that the council was going to demolish our area. We were shocked by this. I remember Mum frequently going up to the Town Hall, and having a go at the planners. It seemed logical to become a planner myself.

The university course I entered the late 1960s bore no relationship to my previous experiences: its content and priorities were alien to me. The feeling was no doubt mutual, as women students who had opinions were a source of amusement or threat. In contrast, the keen young male students appeared to know exactly what town planning was really about, and their confidence and views were accepted as normal and correct. Town planning was presented as being about technology, mathematics, computers, motor cars, football pitches, the future, and a strange form of 'left wing' politics which had it in for women, houses, children, religion, shops, cats, bicycles, parks and gardens (and garden cities), and other such domestic, 'bourgeois trivia'. All this was an intense shock to me, making me feel as if all my childhood and teenage memories were invalid, irrelevant, or had never happened (or had to be 'unlearned', a theme highlighted by Steinem, (1992: chapter 3), whose observations are similar for American women).

IS THERE STILL A PROBLEM?

Nowadays, many would argue that everything is, obviously, better. Approximately 30 per cent of planning students (Table 1, p.194), and over 16 per cent of members of the Royal Town Planning Institute are now women (Nadin and Jones, 1990) (Table 2, p.195). Planning education has expanded and diversified and is less elitist, and even ethnic minorities (who are inevitably seen as male) are not, apparently, doing too badly either (Ahmed, 1989; TCPA, 1990). Planners continue to pride themselves on being concerned with the working class (Greed, 1993a: chapter 11). Whether they perceive this group as anything more than an abstract image, in male workerist terms, as cardboard cut-out people, and whether it includes women is open to debate (Bellos, 1988: 102). Although there are more women in planning quantitatively, this does not automatically mean the situation is better qualitatively (Greed, 1988; 1993b). Many women planners still encounter incidences of thoughtless sexism and discrimination, even from colleagues who boast they are new men. A new generation has grown up over the last twenty years, and although there are problems, feminism is taken for granted.

Compared to surveying (Greed, 1991) and architecture (Lorenz, 1990) there has always been a slightly greater number of women in the RTPI, and in satellite, voluntary organizations (such as the Town and Country Planning Association) concerned with urban communities, housing, and environmental issues and planning (Hardy, 1991). A few women have made an impact on the planning profession, such as Henrietta Barnett, the first woman admitted in 1914, and the builder of Hampstead Garden Suburb (Darley, 1978: 187). Most of the women famous in town planning circles, such as Evelyn Sharp, Ruth Glass, Elizabeth Scott, Elizabeth Denby, Lady Denington and Jane Jacobs have not been members of the Institute. It is only in the last ten to fifteen years that women have gained a small measure of visibility within the Institute. The profession is still male-dominated in philosophy, organization (especially at the top), and policy orientation, marginalizing women and their needs within the built environment.

Less than 5 per cent of senior planning officers are women, and there are policy areas which have hardly been touched by a gender perspective. But the rumour seems to have been spread that the problem has now been solved. It would seem that at some point several years ago the problem was acknowledged by planners. This was equated with it being solved. Therefore the implication is that we should no longer complain, as everything is better now. I get a similar story from some of my students who assure me that feminism was something which happened in the distant past, i.e. the 1960s. I have been told by grown-up planners that there is no need for this book; and by students, 'we've done women' and 'we've heard it all before' but, significantly, 'should we learn it, is there going to be an exam question on it?'. The same criteria of repetition and reiteration of first principles are not applied to topics such as tranportation, conservation or the inner city, which we have heard about for many years, because, it is argued, 'there is always a new generation coming along and it's all new to them'. In contrast, some men achieve brilliant careers on the basis of endless repetition of planning dogma. The purpose of this exercise may not be to impart knowledge at all, its real purpose being to perform an anthropological function – namely through repetition to reify the values of the tribe to initiates (Mead, 1966). Young women town planners may not look to feminism for a solution to their problems, because 'women and planning' already exists but does not appear to offer them salvation.

Over the last ten to fifteen years, an urban feminist movement has developed. But it should not be assumed that the women and planning movement or all its members are feminist. The movement has been accompanied by a wider academic critique of women's disadvantaged position in the city of man (Little *et al.*, 1988; Mackenzie, 1989; Roberts, 1991; WGSG, 1984; Wilson, 1991; Appendix II[3]). Such material has emanated from the 'geography and gender' group of the Institute of British Geographers, and from women in architecture and housing, as well as women planners. You can now do your dissertation on women and planning, or teach it as a course. But some women hesitate to do so, for fear of being marginalized if their main academic work is only on women. It is seen as a bad career move. In London boroughs, where around 29 per cent of all women planners are working, women's progress is marked and there are a few women chief officers (Nadin and Jones, 1990; LGMB, 1992). The work of the former GLC (Greater London Council) women's committee and of women planners, combined with the support of some of the GLC's political leaders led to a series of reports being produced in the mid-1980s (GLC, 1985, 1986a, b, c; the latter two linking women and race). *Changing Places* (GLC, 1986a) is virtually a mini-textbook on the topic, whilst the RTPI's *Planning for Choice and Opportunity* produced in 1989 set out many of the key issues.

But little policy change is observed in the built environment. Transportation planning, in particular, still remains virtually inviolate to women's influence. I see new, out-of-town developments around Bristol and London where there is no evidence of a gender perspective being brought to bear on the estate design, or on the inter-relationships between the land uses at a city-wide level. But excessive publicity is given to the provision of a single crèche, and to small changes such as dropped kerbs. Gender-biased attitudes are evident at the highest levels, such as when the Transport Minister Roger Freeman suggested that 'typists might travel cheap and cheerful' on a second-rate commuter train service to London, appearing to imply they were causing overcrowding and taking men's places (*London Evening Standard*, 13 January 1992: 3).

Whilst everyone pretends women and planning is an important issue, interest fluctuates. In Britain, planning is primarily a local government function and 80 per cent of all planners (but nearer 70 per cent of women planners) are found in the public sector. During the 1980s many local authorities appointed Equal Opportunity officers (EOC, 1988), and one might find a woman, usually *one* woman, appointed as women's co-ordinator or planning equal opportunities officer to advise and monitor policy in planning authorities. But, such a person might be appointed on a junior grade, and be expected – all on her own – to make sure the men toe the line (Foulsham, 1990: 244–58). As one such woman commented to me, 'do they seriously expect me to butt into the conversations of huddles of senior men, and tell them, "look here chaps, you're doing it wrong"?; they would laugh me out of the office'. Such responsibilities put considerable strain on the incumbents of these posts. Women planners have observed that such initiatives have been 'offered without timescale, budgets, or priorities' and have been labelled as, 'simplistic, inadequate, and tokenistic' (WEB, 1990). Whilst many local authorities have EO (Equal Opportunities) statements, few are translated into codes setting performance requirements: there is no enforceable BS (British Standard), such as BS5750 on 'Practice Management Standards' (nor the equivalent of the 1S0 9001 International Standard, and EN 29001 European Standard on management performance) nor QA (quality assurance). There are no monitoring requirements such as govern other aspects of the built environment professional's work. Nevertheless many of the older women who entered planning in the 1960s and 1970s are still making the running in seeking to bring equal opportunities to local planning authorities, and some, at least, are being admitted into more influential management posts. But, gender segmentation patterns are evident in town planning employment with women going in at lower levels than the men, often on part-time or short-term contracts (chapter 2). In spite of EO statements, employers are managing to discriminate (Collinson et al., 1990), pleading 'lack of resources'.

Women have scarcely had time to complain about such staffing policies before they have found themselves being overtaken by changes in the political structure of local government (which have generally moved right), and by reductions in its funding (Aisbett, 1990), all of which has led to the axing of some posts in women and planning. These changes were undoubtedly precipitated by the changing economic situation as the property market moved from boom to recession, making it necessary for both public and private sectors to make cuts. Women were seen as easy targets, for those who were 'last in' were likely to be 'first out'. This put those feminist councillors who were leaders of London Boroughs in an impossible position (Coote and Pattullo, 1990: 23 et seq.), and reduced the likelihood of women-centred town planning policies. As a result of greater emphasis being given by local government in the 1980s to women and their needs, and the support given by some senior officers, women-related policy statements had been written into some policy documents such as the new Unitary Development Plans (UDPs) under the 1985 Local Government Act. This act also abolished the GLC on 1 April 1986, creating a feminist fall-out which spread to London boroughs resulting in women related policies appearing in many of the borough UDPs.

The situation today is very patchy. Some authorities have introduced women-related policies whereas others have scarcely heard of women's issues, or see it all as a joke: rural districts being seen as 'the worst' (Halford, 1987). Women who are highly respected nationally in the women and planning movement sometimes have to fight with their so-called equal opportunities employers to be allowed time to speak at conferences and then are expected to take this time as unpaid leave. The British planning system is operated primarily at the local government level with central government through the D.o.E (Department of the Environment) giving policy guidance. There has never been a D.o.E policy document on women and planning, whilst sport gets a whole Planning Policy Guidance note (PPG) to itself (D.o.E, 1992c). An early draft of PPG 12, Section 5.48 (D.o.E, 1992a) included the word 'women' but this was later removed. Also the D.o.E (1992b) *Development Plans: Good Practice Guide* states, within a discussion of which social issues and 'particular sections of the population' should be considered in the plan-making process, 'perhaps, children, women and homeless people should be added to the list' (para. 3.75). It recommends that it is inadvisable for there to be separate chapters on women and ethnic minorities (para. 3.79), preferring reference to such groups to be integrated (and lost?) within mainstream chapters in policy documents.

The dust is settling on what was achieved in the 1980s, and we have entered what many see as the resource-starved, recession-ridden, post-feminist 1990s. Some of the key figures in the women and planning movement of the 1970s and 1980s have left local government. Some have been made redundant when the funding for their women's unit ran out, whilst others have moved on to different employment, either out of sheer exhaustion and frustration, or in order to give more attention to their families. Local authorities may prefer to employ and promote those women planners who are seen as moderate (but who are still women and therefore satisfy EO policy), such as the the patriarchal woman and the femocrat (woman bureaucrat) (Watson, 1990; Dale and Foster, 1986). Such women's success may often be attributed more to personal opportunism than to equality policy, or be seen as the accidental result of managerial decisions which are not related to the objectives of the women and planning movement. More radical types of women may be seen as too threatening and unsettling.

CONTEXT AND INTER-RELATIONSHIPS

In order to read the game one needs to understand the relationship between town planning and feminism as two of the most important social movements which have developed in modern times. Feminism is not a unitary movement, there are many feminisms and some varieties are more amenable to being linked to town planning than others. Feminism can act as a prism through which one can evaluate the built environment. It can serve as the driving force for seeking to change it, as manifested in the current women and planning movement. But, the priorities within certain types of feminism can detract from the importance of the built environment. One of the reasons I am writing this book is that it might only take a few years for the whole women and planning movement to be rendered invisible, and thus be buried for posterity. I purposely name individual women planners to right the balance, as 99 per cent of the planners usually mentioned in the written record are men. There was an active first-wave women's town planning movement at the turn of the century has taken many years to unearth (Hayden, 1981; Pearson, 1988). It was lost partly because of its exclusion from the official story of planning, but also because of the particular emphases prevalent in the second wave of feminism, many of whose proponents were from arts or humanities back-grounds for whom spatial and technological issues were of limited interest.

The first wave of feminism, with its emphasis on material issues, had links with the early town planning movement. The second wave of feminism from the 1960s seemed to develop quite separately from town planning, until women inspired by the feminist movement applied the principles to the urban situation and created the women and planning movement. The first wave came *before* the establishment of the governmental planning system, whereas one of the chief problems of the second wave of urban feminism was the need to contend with an *existing* state planning system. Although many urban feminists may see town planning as a tough adversary, as will be illustrated, it is quite a fragile and fluid phenomenon. In spite of spurious outward images of technical hardness and spatial fetishism, planning has one of the most volatile academic discourses and range of college syllabuses, and one of the most variable professional cultures of all the property and land use professions (in comparison, surveying is relatively mono-dimensional, Greed, 1991: 37). Planning can be anything you want it to be. It is a container into which one can deposit all sorts of topics. Because of deficencies in social welfare provision in Britain, women planners have sought to use planning in a compensatory role to achieve better childcare, for example, by seeking to extract crèche provision from developers as part of planning gain (pp.184–6)[4], whereas in other European countries with a higher level of childcare provision women planners can devote more of their energies to other issues.

One has to be cognisant of wider social, economic and political changes to understand why feminism came to prominence when it did, and why and when it was allowed to enter the halls of town planning. In recent years there have been some major changes in the construction of feminism, and a mood of disbelief is setting in (Fraser and Nicholson, 1990). Many women were so pleased to have discovered feminism as an explicit ideology, or code of faith, which explained all the things about their lives which had always worried them, but for which there was no name (compare, Friedan, 1963: 13–29) that few were willing to express reservations, either because they did not want to appear disloyal to feminist hegemony, or because they thought it was just them. Adopting a feminist and a women-and-planning perspective allows room for some of my misgivings about town planning, but it does not accommodate all my un-ease. With the expansion of higher education, more people who are working-class, older, and/or female enter planning courses, and criticisms rise.

In studying women and planning one is inevitably drawn into the study of the story of town planning itself, into investigating the changing nature, scope, philosophy and culture of town planning, the priorities which it has manifested, and the varying levels of access to the planning system given to different types of people, both as planners and planned. Pursuing lost feminist issues becomes a gateway to finding other unexpected components within the town planning movement. There was a strong religious and visionary dimension to historical town planning which is excluded from modern humanistic planning. A diverse range of elements which were components of the early movement were expunged from the official record, including eugenics, spiritism, temperance, homoeroticism, evangelicalism, co-operative housekeeping, theosophy, kabalistic geometry and colonialism (as explored in later chapters). It is fascinating to ask what has been left in and what has been excluded from the discourse of town planning and why. Planning was fuller in scope and more diverse in content, more qualitatively orientated, and culturally richer in the past. One can observe a gradual narrowing down of its remit to a concern with technical, physical and quantifiable issues which could be accommodated within bureaucratic planning structures.

THE URBAN QUESTION

In writing the story of women and planning, I am seeking to provide, from a feminist perspective, a small part of the answer to the urban question, 'who gets what, where, why and how' (Pinch, 1985). This is a sequel to my study *Surveying Sisters* on the role of chartered surveyors (Greed, 1991) with which I will inevitably make comparisions. I will seek to cast light on the planners' part in the reproduction over space (and time) of social relations, that is the imprint of gender relations on the built environment (Massey, 1984: 16). One of the problems in trying to disentangle the precise effect that surveyors have in this process was that 80 per cent of them work for the private sector, and therefore they may be seen as the agents of capitalism as well as patriarchy. Town planners are meant to be the 'good guys' the compensatory profession seeking to limit the worst effects of the transmission of capitalism onto space, so one ought to be able to see and isolate the patriarchal elements more clearly.

I found the surveying subculture and related world view to be relatively narrow; surveyors are united by a desire to please the client, and get the best financial return from the site, and have little interest in the wider implications of the development (cf. Joseph, 1978). Town planning is quite different and much more complicated; its subculture is more fuzzy around the edges and changeable over time. Town planning is more of a movement than a profession. Certainly there is no surveying movement as such. Planners are not meant to be motivated by profit, but more by ideological principles. Motivations vary from a genuine desire to make the world a better place by means of reform, to a longing for power and control in order to impose a particular version of order and reality on society. Oddly, the nature of the ideology promoted does not seem to matter too much, provided it legitimates the planners' right to intervene. But the urban end product remains remarkably similar – the same land-use policies are promoted – although the reasons for them may vary.

THEORETICAL BASIS

The purpose of this book is to show the fundamental gender bias in the philosophy and practice of town planning, to investigate why and how this came about, and to consider the potential for change. Therefore, it is implied, as a starting point at least, that patriarchy is the problem. It is not the place here to prove *de novo* that patriarchy exists, and that feminist theory is valid, as this has been done in many books over the last twenty to thirty years. The reader unfamiliar with the basics is directed to Appendix II, p.197, where a list of primary, explanatory texts is provided. This text is not presented as an argument to prove the theory of patriarchy, rather I identify its various manifestations, to illustrate the problem with which the women and planning movement has to contend. But I seek to take the discussion on further, in investigating why patriarchy manifests itself in the particular way it does in the structuring of the built environment, and in considering what other factors contribute to dissatisfaction. Indeed, in these post-modernist times, the validity of single causal structural theory is being questioned, be it based on gender or class. 'Belief' (how a person 'sees' the world) is a primary factor in understanding why people of the same class or gender have different life experiences in the same physical space; they inhabit different social and ideological space. Diversity of life styles, personal perspectives, ethnicity, state of health and age, and subcultural allegiance within and across classes need to be appreciated in understanding a particular person's experience of urban life, and thus their expectations of planning policy (Healey, 1992).

As a town planner concerned with land uses, developments and urban issues, I gravitate instinctively towards a topic-related, rather than theory-based approach. As Harvey (1975: 24) implies, we do not exist in a spaceless vacuum, but live in a material world. However, abstract theory can play an illuminating and sensitizing role (Cooke, 1987). My own theoretical position

9

might be described as feminist neo-weberian, seeing neither patriarchy nor capitalism as ultimate causes, but as primary manifestations of deeper power structures and societal forces (Weber, 1964). These forces show themselves, *inter alia*, in discriminatory attitudes by dominant groups towards a range of subject groups. As to how power, and thus patriarchy, is instrumentalized in society, I see inter-personal relationships as central to the process and gravitate towards symbolic interactionism as an explanatory theory. This theory may be seen to be not just about being concerned with individuals' roles, but as an organizing framework which can be built up into a structural theory of society (Blumer, 1965). The professional subculture of surveyors is the primary vehicle by which patriarchal values are translated onto space. The town planning profession also possesses its own subculture, that is, its own cultural traits, beliefs and lifestyle (Howe, 1990; Estler *et al.*, 1985; Knox, 1988). One of the most important factors seems to be that a person fit into this culture. To achieve this a person's world view – their personal beliefs – must conform to the belief system of the professional subcultural group, the planning tribe. Such are the forces of professional socialization that a woman planner may operate primarily as a professional and not as a woman (Rydin, 1993: 218).

The personal attitudes and beliefs held by members of the subculture have a major influence on the nature of their public, professional decision-making, thus influencing the nature of urban development (Greed, 1991: 6; Howe and Kaufman, 1981). To understand this process one needs to map out the linkages between micro-level personal attitudes and interactions at the private level, and consider how these feed into establishing group identities at the meso-level of the professional subculture. These in turn are the building blocks of societal structures in public life at the macro-level which maintains patriarchy (explained previously in Greed, 1991: chapter 2; Greed, 1992a). Some readers might argue that planners are transmitting onto space other phenonoma in addition to patriarchy, such as capitalism or racism. An emphasis on patriarchy does not preclude these possibilities. Readers who come from other urban disciplines might already have in their possession other pieces of the jigsaw, as they too (like me) seek to put together a picture of the process of the reproduction over space of social relations (Massey, 1984: 16). Planning policy is, of course, closely linked to urban theory deriving from other 'malestream' academic realms such as geography, sociology and economics. Traditionally the problem has been seen in terms of class and capitalism, rather than gender and patriarchy. Planners have often legitimated their power by stating that they are planning for the good of the working class. The construction of class, of what is considered wrong with capitalism, and what is perceived to be the right solution to meet the needs of the worker, are all highly gendered. Debate rages as to the relationship between patriarchy and capitalism, especially as to which has precedence as the causal factor. To appreciate the controversy the reader is recommended to consult McDowell (1986) who argues for a class-based explanation of women's subordination, Hartmann (1981) on the relationship between capitalism and patriarchy, and Walby (1990) on theorizing patriarchy. Other writers feel that neither of the reductionist discourses based around gender or class can contain their life experiences, particularly those from ethnic minority groups and other so-called minorities. Some argue that race should be seen as a primary factor, alongside class and gender, in understanding urban spatial structure (Smith, 1989; Cross and Keith, 1993). I acknowledge the importance of these other perspectives and draw on them too.

BELIEF

When I was studying surveyors I developed a model to understand the reproduction over space of social relations, as to how the imprint of gender relations on the built environment came about (Greed, 1991: 21), encapsulated as follows:

Gender, class → Surveying subculture → Space.

The situation is more complex in planning, but can be reduced to the following thumbnail sketch:

BELIEF → Gender, class → Planning subculture → Space.

Thus the subculture acts a vehicle for transmitting beliefs onto space through plan making. (Surveyors have a belief system too (Greed, 1991), but in planning the link is more overt.) Belief is an elastic word which might be seen as a continuum-concept, containing secular and spiritual elements, the spectrum running from politics/ideology/philosophy, across to morality/religion, and on to spirituality/faith and matters supernatural. In the past town planning was extremely belief-driven, both in respect of ancient, religious city planning, and in the development of ideological, utopian and religious communities in the nineteenth century. The modern, secular planners' world view, and patriarchy, may also be seen as powerful religious systems. It was helpful to adopt a cultural anthropological perspective (Mead, 1949; Ardener, 1978, 1981), to develop an understanding of the social construction of reality (Berger and Luckman, 1972), and the nature of the social meanings and taboos (Foucault, 1976) expressed in the religion of the planning tribe (Eliade, 1959; Durkheim, 1948). Planning is all about creating realities (or rearranging existing reality) and imposing these on space, often obliterating others' true realities and needs in the process (D.o.E, 1972a). In this study I combine material from secular, feminist thought with concepts from religious, feminist sources (ranging from de Beauvoir, 1949, to Heine, 1987, 1988, and across to Goodison, 1990, and Daly, 1993).

I entered town planning espousing another belief system, with a mixed set of class experiences, having lived my teenage years in an inner city, multiethnic area, becoming somewhat '*déclassée*'. This gave me a different personal prism through which to view planning long before I discovered feminism. A year before I began to study town planning, I had been converted from a free-thinker family background to a fundamentalist, pentecostal fellowship (led by two women ministers). I could not help but draw parallels between the two total belief systems (planning and Christianity) which were fighting for my young heart and mind. It seemed to me that town planning had all the characteristics of a religion, however much the proponents of planning denied this and saw me, not themselves, as fanatic. Also, I was puzzled by the way that the spiritual dimension was left out of modern academic discourse. Like Arthur (1991) I have always wondered, 'How can the humanities ignore the human soul?'. As Eliade demonstrated (1959) a characteristic of religions is to make a clear division between the sacred and the profane, between right and wrong, to impose an order on the world and thus create a reality in which man is right and at the centre, thus to give meaning to his existence, and woman is 'other'. This division may be expressed by gendered divisions in urban space, with separate zones for men and women, corresponding to public/private dichotomies within society itself (Ardener, 1978, 1981). How planners 'believe' the world to be is a crucial factor in determining the nature of town planning policy, not least in determining women's place in the city of man. In seeking to expose the dynamics of how patriarchy is reproduced over space, the identification of dichotomies and consideration of their effect on the nature of town planning became a key organizing factor and theoretical framework.

CONCEPTUALIZATION

Many of the scientific divisions which have been promoted by town planning may be understood as manifestations of ancient, patriarchal beliefs about woman's place in the world of men which are expressed as dichotomies and maintained by spatial division, (cf. Douglas, 1984; Armstrong, 1993) – enabling separation to be made between sacred/profane, clean/dirty, and public/private – and thus between man/'other', us/them, professional/personal, majority/minority, same/different, real/unimportant, objective/subjective, work/home, outside/inside, production/consumption, red/green, breadwinner/homemaker, physical/social, economy/culture, spatial/aspatial, insiders/outsiders, public service/trade, order/chaos, adult/child, perfect/faulty, science/nature, (but also art/science), town/country, elite/masses, bourgeois/proletarian, planned/unplanned, rationality/emotion, normal/exotic, white/black, past/present, future/present, new, young/old, spiritual/material, mind/body, good/bad, evil, professional/academic, professional/trade, bureaucratic/entrepreneurial, visible/invisible, serious/silly, middle-class/working-class, acceptable/unacceptable women.

The use of dichotomies in feminist research is a respected practice and theoretical tool (cf. Oakley, 1992: xi). But, as de Beauvoir commented, from an existentialist feminist perspective, identifying the divisions does not explain why women usually end up on the wrong side of these divisions (quoted in Rosser, 1992: 543). The situation is complex, as sometimes the two sides of a dichotomy seem to interrelate and overlap in daily life (Lewis, 1992a) and may not be hierarchical, but just different (cf. Bryson, 1992: 173–6). Nor are they immutable, for, as will be seen, the side to which an attribute, activity or land use is assigned can change. Since these divisions are clearly not natural (Vaiou, 1992) and not everyone may accept the need for them, they have to be actively imposed and maintained, even exaggerated, to gain credibility. Town planners may be seen as the agents of this process, as 'zoners': that is, those who assign and divide space to reflect these beliefs. In contrast I identify types of women from the historical record who have acted as 'zone zappers', who have challenged, and trespassed across, the public/private divide, thus moving vertically from the underside to the overside of public life (Boulding, 1992), creating new space for women in the process (Wekerle et al., 1980). Members of the women and planning movement continue this tradition today.

METHODOLOGY AND TONE

I did not undertake a detailed ethnographic study (see, for example, Hammersley and Atkinson, 1983) of planning education or practice, as in my study of surveyors. However, there is a small element of ethnographic anecdote added where appropriate, whenever someone came up with a *bon mot* which encapsulated the situation, or women planners shared their personal experiences, or events occurred which proved sensitizing. Male/female dichotomies are particularly visible when patriarchal, public beliefs are expressed in personal comments, actions, and attitudes towards women in the private realm. Rather because town planning is such an ideological activity, I explore the overarching issue of belief. I use a qualitative *verstehen* approach (Weber, 1946) as I seek to understand and make sense of the situation, by unpacking the social construction of the planning tribe's sense of reality (Berger and Luckman, 1972). My study is based, to a considerable degree, on written sources. In particular, in chapters 6–9, I draw illustrative material from *The Planner*, the journal of the RTPI and the main journal for town planners in local government, as one mirror which reflects planners' beliefs.[5] I also draw on a range of other academic and professional material. I use a broadly chronological framework to illustrate the gendered nature of planners' and urban theorists' beliefs as to how the world ought to be. My aim (for example in chapter 5, which is on the history of planning prior to the nineteenth century) is not to produce a comprehensive, maybe over-ambitious world history, but rather to present a small selection of examples of how others, including women and ethnic minority students, might perceive the history of planning. I seek to point out public/private dichotomies latent in the planning discourse, and to present 'alternative' connections, juxtapositions and alignments, thus 'making the familiar strange' (Delamont, 1985). In most academic work, let alone professional town planning literature, 'man' is at the centre of the known universe and 'woman' is the 'other' (Haraway, 1989, 1991). I use the word 'other' as an allusion to the perspective of those (including me) who are looking in from outside, who are excluded or invisible. I seek to combine the use of written public material with personal observations of the contemporary world of town planning from an 'other' perspective.

If a generalized observation is included this is because it is something which I have noted many times, not a one-off instance. With some illustrations I have been intentionally non-specific as to sources because it would be unwise (legally and ethically) to cite names, dates, or places. This uneasiness and intentional imprecision may be a feature of feminist research, in which one feels something must be said, to bear witness as a form of resistance, but at the same time one feels vulnerable. I occasionally 'see-saw' (in style and content) between dealing with material from conventional academic and professional sources, to the inclusion of reflections, experiences, and opinions from the personal realm, to emphasize public/private dichotomies. I often preface a see-saw in the text by the phrase, 'it seems to me', although the book as a whole is written in the personal voice. Personal interpretations and experiences are important for, 'in this process I too am subject' (Mulford, 1986), as both planner and planned. Methodologically, this might be seen by some as anecdotal but I consider such elements are valid as illustrative material of the dichotomies under discussion, as based on life experience (cf. Stanley and Wise, 1983, 1993 in which this issue is developed at length). Nor should this material from the personal realm be seen as marginal, intrusive, irritating or 'frivolous' (Rose, 1993: 9) interrupting the flow of the book, but as central to the debate in showing the dichotomies and contradictions between the public and private realms in planners' lives. Feminists have long argued that the 'personal is political' and, by extension, the personal realm is of central spatial importance in urban feminist research, of which this book is one manifestation, both in content and style. Therefore the topics and illustrations included should not be seen as marginal, incidental or 'limited' or 'only' relating to women (because they relate to the personal or domestic realm); from a feminist perspective they are seen as central, and encompass the whole of urban society, relating to men as well as women.

Where I include personal recollections, I cannot remain retrospectively ingénue, and so seek to make sense of past experiences as to 'where I was coming from' and 'where the planners were going' in respect of gender, class, and belief. It is important to 'leave the researcher in' (Greed, 1990), for how else can one know what personal baggage the town planning writer brings to the task? I invite the reader to use his/her life experience to judge my evaluation of how planners see the real world, and my perceptions of what constitutes the real 'real world'. I was unable to make an extensive range of visits, but I was frequently in contact with planning colleagues in the course of my normal duties. Following faculty reorganization I transferred from the Department of Surveying to the School of Planning, which gave me more contact with planning lecturers, students, and practising planners. Attending town planning conferences, and sitting on various committees, including the RTPI national women's panel, provided gateways to a range of written sources and a variety of women and men in the profession. Also, I wrote to all planning schools, to regional branches of the RTPI, and to local authorities, to obtain quantitative contextualization material, and to elicit views on 'women and planning'.

It will be readily apparent to the reader that I am engaging in controversy, that I am angry about the present nature of the built environment, about missed opportunities, blind spots, gendered perceptions and the misuse of power which are associated with the nature of the planning system, and the planners who operate it. Whilst *Surveying Sisters* was relatively restrained and subtle in its presentation of a feminist perspective, the reader will find this book more direct, opinionated and personal in approach. This book should be seen as a polemic in the tradition of Jane Jacobs, who started her most famous book with the sentence, 'This book is an attack on current city planning and rebuilding' (Jacobs, 1964: 13). Planners are not entirely to blame, because others besides the planners have a role in shaping the built environment, including private-sector developers, other property professions, politicians and urban theorists. Chapter 2 discusses the scope and nature of the planning system, because planners' power is circumscribed by legislative and administrative factors. In chapter 3 I will discuss the problems generated for women by the man-made built environment, but firstly it is necessary to discuss 'who' we are talking about. Neither men nor women are unitary groups, and thus I reflect, from a critical feminist perspective, on the definition and application of the concepts of gender, class, man, and feminism. In chapter 4 I step back from town planning and discuss 'planning' as a phenomenon from a feminist perspective. Chapters 5–8 illustrate how the present situation came about, with reference to the historical and modern development of town planning, whilst chapter 9 highlights urban sociological perceptions of women which have influenced planning policy. Chapters 10–11 show how the situation might be changed with reference to the initiatives of the 'women and planning' movement.

UNRESOLVED ISSUES

There is another set of dichotomies which will be found throughout the book in respect to the implementation of planning policy in general, and equal opportunities issues in particular: theory/practice, spirit/doctrine, idealism/bureaucracy, words/deeds (especially), public policy/personal experience, image/reality, idealism/pragmatism, expectations/disillusionment. In respect of illusions of progress, and the little achieved, one must also be on the look-out for the operation of those false antitheses which sanction 'benign' discrimination, such as equality/difference (not equality/inequality, note). In local authority planning these dualisms are not easily bridged, although, practical women and planning policies can reduce the distance between grand theory and practice. This book is not intended to be a prescriptive textbook on how to plan for women, but specific policy statement documents and 'design guides' (such as GLC, 1986a, WDS, 1990, Southampton, 1991, LWPG, 1991) will be considered in the last chapters within the context of resolving dualisms.[6] Those readers unfamiliar with town planning should consult my introductory textbook (Greed, 1993a). Whilst I can show what is wrong, and can offer policy examples of what sort of built environment could exist if gender considerations were taken into account, there is no instant answer to the question of how to put theory into practice, or how to turn resistance into resolution. To change the spatial situation one first has to change aspatial (social) structures (Foley, 1964: 37). The organization of the planning system must change if more women are to achieve positions of seniority and power in order to be able to put policy into practice. Many ordinary people, especially women, have little hope of being seen as suitable material to become planners by the gatekeepers – the two-faced Januses – guarding the portals of the profession (Spender and Spender, 1983). Thus alternative ideas from outsiders, and 'others' are unlikely to form the basis of policy on the built environment. How we can change the duplicitous belief system underlying the nature of the planning profession is a daunting problem. We are, after all, threatening the very power structures which project patriarchal beliefs on urban space.

NOTES

1 Roberts, 1991: ix. Roberts describes many women like myself who, on entering into built environment education in the 1960s and 1970s, felt a huge culture shock because of the maleness of the courses. But, many young women of the post-war baby boom generation in general were no strangers to the feelings of disorientation, disappointment and confusion brought on by their experiences of the so-called education they had endured at all-girls schools, as described by Ingham (1981) and Heron (1985). Neither extreme was ideal.

2 My thanks to Caren Prys Jones, Information Officer, Port Sunlight Heritage Village Historical Collection, for digging out copies of the original Rinso packets, and for her own insights.

3 References which explain, expand, or give background on the topic under discussion are included within the text, clusters providing signposting to related work. Page references are given for specific points but these are not elaborated upon, or justified, by the provision of footnotes. In a text of this comprehensive nature the aim is to provide immediate, same-page notification of sources for the reader. Appendix II comprises categorized primary text lists.

4 Planning gain is a benefit gained for the community, which must be site-related, such as the provision of landscaping or public amenities in return for a more favourable planning permission.

5 Journal references in the text (mainly in chapters 7–10) refer to *The Planner* unless otherwise indicated.

6 Further information on 'how to plan for women' is given in Greed, 1993a, chapter 12, which constitutes my introductory 'mainstream' planning textbook. In the present text, Appendix II (p.197) provides an introductory summary of 'women and planning' texts for the newcomer. Appendix III (p.201) sets out a summary of policy proposals and initiatives.

2

THE PLANNERS
Powers and limitations

BLAMING THE PLANNERS

In this chapter the context of the planning system and the representation of women therein will be introduced (Greed, 1993a: Part I gives a fuller account). It is easy to blame the planners for the gendered nature of towns and cities, for if we assume a direct link between the nature of the built environment and town planning, this implies that the planners have unlimited power. But planners are only one set of actors within the development process which creates the built environment. Others include private-sector developers and their professional advisers, politicians, and, more broadly, urban theorists, other urban economic and social policy makers, and a variety of cultural and entrepreneurial trend setters. The nature of the planning system is problematic at source, because of its legal scope and limited powers, rather than because of 'who' the planners are. Even if more chief planning officers were women it might not follow that cities would be better (cf. Greed, 1988), or that planning policies would be different. One must consider the nature and organization of the planning system, the limitations it offers to change, and the types of women and men who are attracted to, and accepted by, the planning subculture. The efficacy of the planning system in addressing women's needs is influenced by the style of management, policy priorities, level of political support and professional perspective adopted by the planner, over and above the legal requirements of 'his' job. The imprint of gender relations on space is not a mechanistic process, and is more likely to be achieved through the spread of ideas, and visions than through enforcement of planning policy.

The planning system can effect relatively little change in the city in any generation, short of knocking it all down. Only a small proportion of the built environment is redeveloped each year, (D.o.E, 1992d). To a considerable degree (especially in historic European cities) the nature of urban form has been established by past generations (see chapter 5). Powerful historical, cultural and ideological forces are at work determining what society thinks cities should be like (compare, Foucault, 1972, on the power of the archaeology of inherited wisdom and knowledge). As Reade (1987: 161) explains planners

18

may have delusions of grandeur. Some sociologists have been taken in by this. There is a range of publications in which planners, and architects, are presented as powerful beings, creatures obsessed with a sense of professionalism, or power-mad bureaucrats promoting state control (Simmie, 1974, 1981; Kirk, 1980; Dunleavy, 1980; Malpass, 1975; Bailey, 1975). None of these generalized images bears much relationship to most of the human beings called planners working in planning offices. Reade (1987) infers that planners are in reality fairly weak, low status beings, but they keep going in the hope that their humble situation is temporary and 'after the revolution' they will be appreciated and start 'real planning'. The hard, no-nonsense professional front (Goffman, 1969) that planning presents to the public, belies a complex internal set of paradoxes, professional squabbles, and unresolved dualisms.

OTHER ACTORS IN THE DEVELOPMENT PROCESS

The developers

The local authority town planners are not necessarily in a strong position to get their policies implemented. They may not initiate policy but reflect in their plans the influence of external ideological and economic trends, and theories. The planners may be chiefly transmitters of social relations. Likewise, the surveyors reflect the values of their capitalist masters in their professional decision-making (Greed, 1991: 21). Alternatively, they may be more interventionist, acting like plugs and sluices in the system, stopping the downward transmission of existing social relations and replanning the flow. In Britain, nowadays, the main initiative for development comes from private-sector interests whose aim is, primarily, to make money rather than to deal with social need or create a beautiful environment. Property is a commodity like any other (Rydin, 1993). There are a range of property and construction professionals who act as advisers and experts to the private-sector property developers (Greed, 1992a: 24–5; Ambrose, 1986: 68–9). The construction industry employs over 2 million people (Dolan, 1979; Ball, 1988; CISC, 1992a), and over 95 per cent of these are male (Greed, 1993a: chapter 3). I consider that the role of the chartered surveyor is central to the development process, because they give advice to the private sector on all aspects of land use and development; the RICS (Royal Institution of Chartered Surveyors) is one of the largest professional bodies (Appendix I, table 1) (Greed, 1991). Nevertheless, I believe that planners have influence as great as that of surveyors, even greater in some instances, but in a more compensatory or remedial role. They act as regulators rather than initiators of development, seeking to control the effects of the private sector.

It should not be assumed that the planners' role is primarily negative. The question of whether 'the market leads the planners or the planners lead the market' is one of the old chestnuts of the property world (cf. Balchin and Bull, 1987). The planning system may be the subject of extreme pressure, with planners' resistance being worn down by multiple applications and expensive planning appeals. But in high-land value areas, where developers are desperate to get permission, planners may have a more positive, pro-active role in directing development, than in an economically depressed area where the local authority would welcome any development which provided jobs and a source of rating income (property taxation). It is no coincidence that the London boroughs have achieved the greatest success in getting women's issues on the development agenda. The planners' remit is, potentially, broader than that of the surveyors: to plan for society and the whole of the city, not just to get the best return for an individual client from a particular site. Planners have to consider the inter-relationship between one land use and another, the social implications, and the traffic generated by a particular development. When their policies are backed up by political support planners can have considerable influence in shaping urban development: their power varying both locationally and historically.

The Royal Town Planning Institute is a small professional body – albeit with BIG ambitions – containing around 17,000 members (Appendix I, table 1). Although only 5 per cent of surveyors but 18 per cent of planners are women, there are only half the number of women planners as there are women surveyors (around 3,000 as against 6,000) because there are over 86,000 surveyors. See-sawing: one can get so caught up in town planning, feminism and/or urban sociology that one overestimates their importance. For comparison, it is sobering to read that in the United States in 1982, 50,000 women held Ph.D.s in science and engineering as against 20,000 in humanities. Women constitute a quarter of humanities doctoral students as against 12 per cent of science but the latter field is so much larger, (Kirkup and Keller, 1992; Greed, 1993b). Relatively few surveyors are directly concerned with town planning, so the planning and development division of the RICS is one of the smallest (Greed, 1991: 200), containing around 2,000 members of whom 10 per cent are women. But all surveyors undoubtedly have an enormous influence on 'what is built' because of the commercial perspective which shapes all aspects of their professional decision-making on behalf of the private sector (cf. Ambrose and Colenutt, 1979). Incidentally, the only woman (so far) to be elected president of any of the RICS divisions was in planning and development – Mary Dent, executive director of planning and conservation, Royal Borough of Kensington and Chelsea from 1989 until her death, at 51, in September 1993. Planners exist throughout the world as a powerful interest group. The British planning profession possibly has more in common with its international counterparts (Rodriguez-Bachiller, 1988; Ball, 1985) than surveyors do because the latter constitute an especially English profession (Thompson, 1968; Greed, 1992b: 211).

The politicians

Women are often overwhelmed by what appears to be the powerful nature of town planning, but members of other property professions might judge the legislative powers of the planners as not constituting a particularly serious threat to their activities. I have heard the town planning profession described as a fragile house of cards which could easily be toppled, because of its need for political support to survive. Town planning in Britain is only as powerful as the government system which promotes it. It depends for its existence on the continuance of the acceptance of the need for state intervention in the built environment. In the post-war reconstruction period of the late 1940s and under the Labour government of the 1960s town planning was in a stronger position *vis à vis* the private property sector than it is today. Were there not a local government system, town planning as a profession as we know it probably would not exist, and many town planning courses would not run either (Rodriguez-Bachiller, 1988: 60).

Planners are not (meant to be) the ultimate decision-makers but the advisers of their political masters, that is the elected councillors on the local authority planning committee. There is an ongoing debate (Montgomery and Thornley, 1990; Healey *et al.*, 1988) as to whether the planners are really the servants of the politicians, or whether they exert power in their own right by dint of strength of personality or possession of professional expertise and monopoly. Planners do not operate in isolation as bureaucrats but as part of the wider local government apparatus, arguably as classical *apparatchiks*. This has its own agenda of urban management, social policy and state intervention (Dunleavy, 1980; Dale and Foster, 1986). The relationship between the planners and their political masters is complex. Many have sought to understand how it works in practice (Healey *et al.*, 1988, especially chapters 8, 9, 10). Most councillors on planning committees are men, albeit with some strong individual women in some authorities (Barron *et al.*, 1988). More fundamentally, the ethos of local government is essentially male (as many town-hall feminists have found to their cost, cf. Coote and Pattullo, 1990: chapters 14 and 15; Rogers, 1988). It seems to me that in the planning process anything can be justified (such as sexism) or lost (such as gender considerations) in a sea of report-writing, and endless committee meetings. If the criteria by which equality is judged are gendered in nature, then women and their needs have little chance of being valued: in spite of a written Equal Opportunities policy statement.

The people?

In theory 'planning is for people' (Broady, 1968). Following pressure from grass-roots community groups in the 1960s, the 1971 Town and Country Planning Act introduced an element of public participation in development

21

plan preparation, which still exists (Section 33 of the 1990 Town and Country Planning Act). Although it is often women who are prime movers in community groups set up to contest planning proposals, many feel isolated from, and intimidated by, the planning system. Public participation and consultation might be seen as a compensatory activity, bridging the gap between public and private realms. This is because of the unrepresentative nature of both the elected representatives of the people and the planners themselves: in terms of class, gender and race composition. It may be considered inappropriate that women as the 'other' should be involved in the public realm of politics at all (McDowell and Pringle, 1992: 9–17), individual women reporting that the planners do not take their views seriously (LPAS, 1986a).

There is limited provision for the public to give their views on development plans. Many are concerned about the lack of the right for third-party representation in the case of decisions over major planning schemes, for example for out-of-town shopping mall development. In the two-sided contest between planners and developers, there is little space for the views of members of the public to be included in the debate. This appeared particularly to be the case in the 1980s, when during the enterprise culture of the Thatcher years the emphasis was upon developer-led planning. In the 1990s there has been a gradual return to strategic plan-led planning. There is also a case for people-led planning as Bob Colenutt (1992) has highlighted, but unless women's views are included in the criteria for community-related planning this may not be an improvement. In ostensibly social programmes of urban renewal such as the 'City Challenge' initiative, evaluation of area profiles may lack a gender dimension, because of lack of consultation with women's groups, and thus plans are based upon an incomplete picture of the problems (WDS, 1992).

SCOPE AND NATURE OF THE PLANNING SYSTEM

Limitations

One should not expect miracles from the British planning system unless one can change its parameters (Greed, 1993a: 19–32 for introductory description of its organization). The legislation upon which planning is based specifically gives the planners the brief of controlling land use and development (Section 55 of the 1990 Town and Country Planning Act); not meeting people's social needs or facilitating how they 'uze' land and buildings. I use 'uze' when describing the activity, for example going shopping, rather than the land classification such as retail development, because it conveys the dynamic, rather than static way in which women 'uze' cities, as they move around in their daily lives. Within malestream planning there has always been debate as to whether it should confine itself to purely physical issues or take on social policy issues.

22

Levels

There are two main aspects to the system: the production of development plans showing planning policy and the operation of development control. In Britain there is no national plan but the Secretary of State for the Environment, in theory, has an overall controlling role in setting policy as aided by his professional advisers (and it has always been a 'he' so far). There is no strong regional level of planning although, in the past, regional economic planning has been a major policy area (Balchin and Bull, 1987). Local government is the main level responsible for planning in Britain. In summary the shire (rural) counties each have to produce a Structure Plan, which is a development plan setting out strategic policy. The policy areas covered include housing, industry, offices, open space, minerals, car parking, transportation, retailing, derelict land, and leisure (D.o.E, 1972b; 1991). The nature of the topics, and the fact that they mainly relate to the public rather than domestic sphere, has been challenged by modern urban feminists (*Changing Places*, GLC, 1986a).

It is misleading to blame members of the RTPI alone, because not everyone working in a professional capacity in a planning office is likely to be a member of the professional body. In a team one is likely to find, in addition to RTPI planners, people from a range of academic realms working as 'planners' including statisticians, sociologists, geographers and economists. One will also find people from other professions including architects, surveyors, planning lawyers and civil engineers. Out of this melting pot of inputs comes the development plan. In spite of the diversity of the team, gender issues are unlikely to figure prominently. This is because all the other disciplines have been influenced, to varying degrees, by the 'same' societal patriarchal perspective as manifested in town planning. Strategic policies from the Structure Plan are translated down into more detailed Local Plans at the district planning authority, which is the main implementatory level. Each county is divided on average into four to five districts. In metropolitan authorities (large conurbations) and in the London Boroughs a new type of development plan, the UDP (unitary development plan) is being produced. This combines, in one set of documents, the strategic policy level of the structure plans with the detailed planning policies for specific areas found in the local plans (Greed, 1993a: chapter 2). Planners can produce policies on all aspects of land use and development but many land-use matters are out of their control, with other departments or statutory bodies dealing with the provision of schools, hospitals, and infrastructural services; although the planners may influence their siting.

Law

Districts are responsible for operating the development control system (granting or refusing planning permission for development). Decisions are shaped by national planning law regulations and the structure and local plan policies for the area. Development is defined under Section 55 of the 1990 Town and Country Planning Act, as 'the carrying out of building, engineering, or other operations in, on, over or under land, or the making of any material change in the use of any buildings or other land'. Therefore development includes both new build and change of use of an existing property or piece of land. Uses and activities carried out by women are not necessarily seen as valid land uses and may be considered *ultra vires*, this attitude reflecting imagined dichotomies between the public/private (male/female) realms of human life and the relative importance of each. As will be illustrated, sport has been seen as a physical land use, therefore a legitimate concern of town planning, meriting its own type of land (public open space), whereas childcare has often been seen as merely a private, domestic activity, although it creates demands for buildings and land. As will be seen in chapters 10 and 11, women have made considerable headway in arguing against this viewpoint, and getting women-related policies accepted as spatial and therefore valid. In contrast, planning decisions based on 'green' considerations and the importance of sustainability although somewhat aspatial in nature are seldom seen as *ultra vires*.

To digress, briefly, development control has a life of its own beyond the world of town planning, being part of lawyers' territory, where it seems that questions of right and wrong, logic or justice pale into insignificance relative to matters of precedent and procedural correctness. Appeals are often won on technicalities rather than broad principles of justice. Although a considerable number of women are involved in development control, and it can offer wonderful opportunities for pro-active planning, women practitioners may feel pressurized, or may have been professionally socialized, to leave their personal views at home. As increasing numbers of women are entering the legal profession, there are now many more women planning lawyers (as well as women planners specializing in law), some of whom do not leave their feminism at home. For example, Susan Nott writes on both planning law (Morgan and Nott, 1988) and women's issues (Nott, 1989). I have scoured the main planning law textbooks (such as Heap, 1991; Telling, 1990), statutes, and the *Encyclopaedia of Planning Law* (Grant, 1993), for signs of awareness, positive or negative, of gender issues, and found little, although Grant's planning law textbook (1990) shows some signs of a broader perspective. Much of the material is 'peopleless' and relates to land, development and uses alone. However, one comes across tell-tale traces of underlying views, such as in Victor Moore's planning law textbook (1987 updated 1991: 58) in which he states that irrelevant policies in structure plans which have nothing to do with land-use planning, referring to policies relating to racial and sexual disadvantage, delay the approval process. But he cheerfully states on page 137 that the 1976 Race Relations Act, Section 19A, makes it unlawful for a planning authority to discriminate against a person in carrying out their planning functions, yet misses the application to the policy-making level of planning, presumably because it is imagined that this section only relates to individuals seeking planning permission.

REPRESENTATION OF WOMEN

Vertical distribution

Even if women have different policies and attitudes on urban issues, they are unlikely to be able to implement them, unless they have reached senior decision-making positions, and even then they may find that their intentions are constrained by legal and administrative structures. Less than 5 per cent of those in senior positions are women, but there are several very senior women in the London borough planning authorities (cf. LGMB, 1992; and Appendix I, table 4). Around 12 per cent of women planners are on assistant planner grade and 4 per cent on principle officer grade (RTPI, 1989b). Not everyone is equally welcome into the professional subcultural group. The concept of closure as discussed by Parkin (1979: 89–90) and first developed by Weber (1964: 141–52, 236) in relation to the power of various subgroups protecting their status is still a key theme. Closure is worked out on a day-to-day basis with some people being made to feel welcome and comfortable (Gale, 1989a and b) and others made to feel awkward, unwelcome, and out of place in the public realm of the professions. As in surveying I would argue that one should not see all the 'little' occurrences of everyday life – the encouragements and discouragements – as being trivial or too personal to count as real data. Rather they should be seen as the very building blocks of the whole subcultural structure, as the nature of an organization is maintained by the actions and attitudes of its members (Blumer, 1965). The need for identification with the values of the subculture blocks out the entrance of both people and ideas who are seen as different (or more likely, ideologically unsound in the more left-wing branches of planning). Possibly it does not matter so much if the applicant is of the wrong class (of origin) as it does in surveying, as a few real working-class northern men around helps create the illusion that planners really are at one with the working class. Many planners have commented to me that planning education may be seen as a route for the socially aspiring working-class man. But, the admittance of working-class women, black men, or black women are three other issues.

One can get a rough indication of the numbers of women in each local planning authority, and their level if they have senior named posts, from the *Directory of Official Architecture and Town Planning* (Brett, annual) because female names are prefaced by 'Ms' and male names have no prefix. In the shire county councils and districts there is quite an irregular distribution of women, so that it is difficult to generalize. However, in one authority it was found that 40 per cent of women were on the lowest scale but no men were, and 50 per cent of men were on the highest scale but no women. One is less likely to find such disparity in metropolitan areas, as there is a greater distribution of women along the staircase up to middle-management levels and in some of the better authorities, around 12–15 per cent of women will be found at principal officer grade. Progress is not guaranteed; as with surveyors there is a log jam developing at the lower management level as more women work their way up only to encounter a glass ceiling. Some women find, after taking a career break, that they return to a lower grade than before they left. Also major reorganization of local government managerial structures has occurred and this has created new male, cross-departmental oligarchies which marginalize women who had previously held senior positions in planning, housing, social services and education.

Not only is the number of women in senior posts low, older women planners have failed to reach the levels of seniority to which they consider they should be entitled. Table 2, p.195 shows the build-up of the numbers of women in the profession. According to Nadin and Jones (1990) three-quarters of women members are under 40, whereas men are fairly evenly spread over the age groups with less than a half of men being under 40 (cf. Table 3, p.195 on student numbers). Between the ages of 25 and 34 women make up a third of all members, but between the ages of 40 and 50 they make up 1 in 10, and from 50 only 1 in 20 of the membership. Less than 1 per cent of all planners are disabled, and relatively few are members of the ethnic minorities. It would appear from conversations I have had that black women are in some cases considered more acceptable than black men, who might be seen as threatening. Whilst 40 per cent of women surveyors are concentrated in London, only 20 per cent of women planners are there but this still represents the largest concentration nationally (Nadin and Jones, 1990; Table 4, p.196). Women planners are spread throughout the country, but with greater numbers, relatively speaking, in the conurbations. In my own region of the South West, it would seem that there are ones and twos even in the 'Deep South' in Cornwall and across in the Channel Isles. Some of the most promising manifestations of the women and planning movement are to be found amongst groups of women councillors on planning committees in unlikely rural locations, irrespective of political affiliation. Whilst many men planners see the planning committee (of elected councillors) as a hindrance, women planners may see it as an opportunity for women in the community directly to influence planning without having to go through the exclusionary process of getting formally qualified.

Local government nowadays is subject to cutbacks, so there are unlikely to be many new posts available as has been the case in the private-sector property professions. Local government is arguably less exciting, slower, and more stable than private practice. Some property professionals have commented that people are failures if they go into local government. Promotion may be, literally, a matter of waiting to step into dead men's shoes. The impression is that the local government fraternity appears extremely cautious in promoting 'unsettling', overtly ambitious or bright women, who might rock the boat, but at the same time genuinely wants to attract more women to the lower levels, because of the 'manpower' shortage, even offering flexitime and childcare vouchers. In contrast, in the private sector women may feel more valued, but can be miraculously promoted one year and then rendered redundant the next, because of booms and slumps in the property market. One should not confuse an upturn in the economy which brings more women into the labour force with an increase in equal opportunities.

If a planning decision is contested it will be dealt with by a planning inspector with delegated powers from the Secretary of State for the Environment. The vast majority of planning inspectors are male, and are chosen through the 'old boy' network, although nowadays one will see open advertisements for these posts. They are frequently filled by professional gentlemen of mature years, with backgrounds in planning, surveying or law, who are judged to have suitable experience (apparently this does not require evidence of dealing with the built environment on a day-to-day level as a shopper, mother or community activist). I am informed by a male planning inspector that around 30 out of the 200 full-time inspectors are women, but that a small, but growing number are employed on 36-week contracts which apparently enables them to combine their work with childcare. In Northern Ireland 1 out of the 12 members of the planning inspectorate is female. Although inspectors are meant to judge decisions on the basis of public policy and law it would appear that those of the 'wrong' personal perspective need not apply. A senior woman planner colleague applied for such a post and was turned down. She was convinced it was because she had been a member of CND, as planning inspectors may be required to officiate over inquiries related to nuclear power. The quasi-judicial nature of planning inquiries may put women off. Selectors may be believe they will not have exude enough personal authority if a disturbance occurs, but most planning appeals are settled in writing, or by less confrontational means, not by huge public inquiries. Paradoxically, there have been a few very senior women planning inspectors, such as Gillian Payne and Betty Harran. There are always the one or two who disprove the rule, in the same way that there are two women high-court judges out of eighty-three incumbants (*New Law Journal*, Vol. 146, No.6554: 746–9, 25 May 1992, 'Minorities in the Profession'). There are also a very few women in senior policy-making positions within the D.o.E, (*The Planner*, 14 December 1990: 66) and a couple of women regional directors. For example, Wendy Pound was one of the most senior women in the D.o.E until her premature death from cancer in September 1993.

In addition to the local government system there is a range of *ad hoc* planning bodies dealing with specific policy areas. The New Town Development Corporations were established with a maximum life of thirty years, to manage the planning of these settlements started in the post-war period. The Urban Development Corporations (such as the London Docklands Development Corporation) have been established to deal with urban regeneration programmes, and are, controversially, quite separate in many respects from the local authorities in whose area they are located. There is no right of public participation in the deliberations of many of these bodies. As different issues and problems become topical over the years, or are pressing in particular geographical areas, special policy groups and committees are built up within the various local planning authorities. There is also a tangle of liaison and consultative groups which have been set up to facilitate discussion of issues of common interest among the different governmental bodies. But no *ad hoc* governmental planning body has ever been set up to deal with women and planning issues. The majority of these *ad hoc* bodies appear to be male-dominated. They are composed of nominated notables who are not necessarily democratically chosen, nor professionally qualified, but are seen as 'good chaps'. I have often heard it said over the years that if one is a retired military officer or an elderly captain of industry, one will be welcome (every New Town committee used to have its wing commander). Apparently, the Government simply has not been able to find enough suitably qualified women planners for such posts (Greed, 1991: 142; Hansard, 1986). Rather like the relationship between the elected House of Commons and the hereditary House of Lords, there is always this other level above statutory town planning which seems to maintain its influence even at times of socialist government, reining planning into the control of the Establishment. On a more positive note the RTPI now requires a 25-per-cent representation of women on each of its specialist panels, although the higher-level committees remain predominantly male.

Horizontal distribution

Until relatively recently few women appeared to have much influence in structure plan policy-making. Until the women and planning movement began to have an impact, one would search in vain to find policies in development plans related to childcare, disability or ethnicity, although waste disposal, minerals and sport were seen as quite valid town planning topics. For example, the Surrey Development Plan (1965: 58) stated 'for physically disabled persons the plan envisages the establishment of a residential home' and that's about it. In London especially, women have pro-actively sought to use the opportunity of the introduction of the new Unitary Plan system as a means of bringing about policy changes. To achieve this, women, some in quite junior posts, network together from different local authorities, and act as an effective pressure group, often with the support of councillors, voluntary groups, and members' women's committees, producing 'external' planning reports and pressing for model policy adoption (LWPG, 1991). They are employed planners but much of their success has come about by stepping outside of the planning system, to promote women's policies.

As in surveying, women planners are generally concentrated in certain specialisms, but to a less marked degree, with individual women found in just about every area: albeit often on a low grade (RICS, 1990). Men, too, are spread across a wide range of specialist areas, and although the general public often imagines all planners work in development control less than a third do so (Nadin and Jones, 1990). Women appear to be more noticeably concentrated in development control, social aspects of local planning, and to some extent, countryside planning. In contrast certain areas are strongly male, such as transportation planning, strategic forward planning, and areas of planning related to commercial development and industry, especially when planning is undertaken in partnership with the private sector. There are individual maverick women working in totally male teams, in transportation, in minerals (a very male area and a county planning function) and in other specialized areas, where, against the flow, they have carved out a niche for themselves. Some such women, although on middle-level grades, may be the only holders of specialist information in their department, so men show them respect for fear of being cut off from vital data. Apart from exceptional women, as with surveying (Greed, 1991: 138–41) it would seem three roles are common for women: first, that of being a 'helper'; second, being involved in social and housing issues – such as dealing with 'minorities' including women – rather than having a managerial, design or developmental role; and third, anything to do with 'prettifying', which could range from the woman herself being given a public relations role, to being involved in landscaping control.

TYPES OF PLANNERS

Persona

Over and above the specialism or job in which a planner works, three main types of planner may be identified. Two of the types identified, whom I call the technician-operator and the manager, work within the statutory planning authorities, and one may do, namely the philosopher king. The latter is more likely to be found working outside the planning system, possibly in academia. Although the image of the 'great planner' designing whole cities might go down well in career talks, in reality many practitioners spend much of their lives doing fairly routine work and have little control over their own destinies, let alone those of others. To a degree the three types identified may be seen as extremes to help make sense of a complex situation. An individual modern-day planner may combine aspects of all three to varying degrees. But one can also identify people who do appear to fit exclusively into one category or another.

Philosopher king

Long before the state town planning system was established there had always existed the visionary, the prophet, the genius, the seer, who might be seen as a planner, in the sense that he (and it was nearly always he) was the architect of a social order. This is in the Edwardian tradition of the great male social leader (Brandon, 1991), or in the sense of Plato's philosopher king, who acted as the guide and guardian of society because of his superior intellect. As will be seen, many of the most powerful people in planning (if power is judged by effect on the built environment) were not planners at all in the professional sense, but outsiders: academics, researchers, writers and creators of realities. Their ideas were believed, and thus they influenced the policies of the planners and politicians who operated the system: that is they led, rather than fitted into, the culture. The most powerful man, that is, Superman according to Nietsche (1973), does not follow any role model, he creates himself as a complete cultural package, and others follow. A softer version of this viewpoint is presented by Weber. This is discussed by Bologh (1990), who highlights the importance of 'heroes' for men as role models. As will be seen scientists, philosophers, sociologists, socialists, artists and architects have been among the heroes who have led the town planning tribe at various stages in its evolution as esteemed outsiders. It is no coincidence that Max Hutchinson entitled his policy statement on the place of women in architecture, 'Heroes' when President of the RIBA (*Architects Journal*, 14 February 1990: 5). In comparison, Alonso (1965: 170) noted that the city planning profession, like most adolescents, likes to revolt, to strike a pose, and rapidly adopts and discards heroes. Unfortunately such heroes, like those of the architects are admired not because they are really revolutionary, but because they embody the values of patriarchal society (Coole, 1988). Heroes are likely to come from an elite background where contact with women, working-class people and those from ethnic minorities is limited: the very people for whom they are planning. Women can exert power over the built environment as philosopher queens, but such women may be seen as marginal, low-status outsiders. They may only be tolerated to ensure legitimation of the planners' claim to be a liberal profession: and not revered like the outsider genius or hero.

Technician operator

At the other extreme is the planner who might be called the operator, technician or follower, who runs the planning system in the local planning office, carrying out development control functions at the implementatory level. Some of these planners might actually be planning technicians rather than full professionals. This type of planner probably most approximates to what the general public imagines a planner to be: a town hall bureaucrat, often caricatured as a 'little Hitler' – typically a man. This negative image is encouraged by media presentations of what planners are meant to be like, ranging from the infamous 'Blott on the Landscape' (Tom Sharpe, dramatized for television by Malcolm Bradbury in the late 1980s), to the weekly rantings of the cub reporter sent along to cover planning meetings for the local newspaper. As a result of these negative, and gendered, images, a woman planner may not be accepted by the general public as a real planner at all.

A technician planner is not a creator of new realities, and is not 'meant' to be a dreamer of visions, but is the implementer and enforcer of others' plans. 'He' might be seen as corresponding to Weber's bureaucratic sub-type of the technician who is without ideas or creativity of his own (Bologh, 1990: 119). In daily practice, an operator planner is not necessarily just following rules, but will use personal judgment and professional discretion in carrying out even mundane tasks, and therefore possesses considerable potential power. Many a local authority planner would argue that by the skilful practice of development control they can have far more influence, albeit on a small scale, on shaping the built environment, than dreamers of urban visions. It can be disastrous if such a planner's decisions are not informed by an awareness of women and planning issues at the local plan and design level. Practising planners who are concerned with 'getting the job done' might see visionaries and academics as irrelevant and impractical people who 'have never worked in a planning office'. Likewise Joseph (1978) draws attention to the importance of the concept of practicality in the surveying profession. My own students are suspicious of waffly academic theory, and prefer practical material, judging the value of different course subjects according to the airy-fairy/ nitty-gritty dichotomy. I found in my study of chartered surveyors the emphasis throughout the profession was on a practical rather than academic image. The practical person was perceived as being 'entrepreneurial' in outlook even if some of their professional work was really in all but name academic research. In town planning the same division exists but is more likely to be reflected in the bureaucratic planner being seen as the practical type in contrast to the visionary, philosopher king. However, some of the great visionaries of town planning, such as Ebenezer Howard, were also entre-preneurs, because the only way to achieve their dreams was to build them themselves. Nowadays the philosopher king planner is more likely to be found in academia, and probably involved in research.

Women can get caught in the gaps between these dichotomies, between technician/visionary, entrepreneur/bureaucrat, and academic/professional. In conversation some women have commented to me that the divisions (which they clearly 'see') between the public- and private-sector professional cultures are counter-productive, and inappropriate since they were framed on the basis of male definitions, into which women never quite fit. Significantly more women than men planners, proportionate to their numbers, choose to go into private-sector planning consultancy, where more entrepreneurial surveyor-like attributes are required. Many people have commented that it is relatively easy to go from public service in a local planning authority into private practice, but almost impossible to go back across the entrepreneurial/bureaucratic chasm once tainted by the outside world. This works against alternative perspectives entering the public sector (cf. Greed, 1991: 26 in which I allude to Miller and Swanson, 1958 in respect of women professionals' personal perspectives). The town planner who is also an 'academic woman' (Acker, 1984) or a 'visionary' is a difficult concept for some men professional planners to cope with, and they may opine that she has no place in 'their' planning office. When seeking a promotion or a first-time job, women may be told they are either not technical enough to be a practical planner because their knowledge is too broad (Greed, 1991: 148), or conversely that it is too detailed to be a strategic planner, or they do not have enough experience. Some women report this happening to them in job interviews on consecutive days. To be a chief planning officer (or so some tell me) one needs basic management skills rather than specific technical knowledge. One must be a team leader (a heavily gendered attribute) as one's time is too expensive to waste on details; one's job is to create the culture, inspire confidence and sustain belief. Nevertheless, quite junior women technician planners can play their part as 'zone zappers', quietly having a positive influence on local area development control decisions, possibly because the men do not bother to check what they are doing as they are not considered a threat.

Digressing briefly, I am increasingly coming across women who, in their life histories trespass across that other deeper dichotomy, deeper even than the technician/professional division, between the manual trades and the professions. The Bristol group 'Women and the Manual Trades Group'; the Leeds 'Women in Construction Alliance' group, and the London 'Women and Manual Trades' organization all include individuals in their membership who hold professional and academic qualifications, but may also be master builders, qualified City and Guilds plumbers or carpenters, breaking down centuries of counterproductive brain/brawn; head/hand dichotomies in construction. After all, much of women's work, especially housework, has always combined the two sides of these dichotomies.

The manager

Between the extremes of planner as visionary or technician there is the policy-making planner who runs the system. The booklet *Careers in Town Planning* (RTPI, 1990) states 'planning is necessary' because:

> Land is a limited resource ... for the general benefit of society it has been necessary to find ways to balance the demand for different uses of land, to locate land uses so they are sensibly related to each other and to try to ensure that land is not developed in such a way that it spoils the environment. Land use planning has developed to meet the social and economic needs of people.

Other definitions of planning include bringing forth land for development, reconciling conflicting demands, economical use of scarce resources, and providing for non-profit-making uses and infrastructure. In these definitions the planner is not only presented as a policy-maker, but as mediator between conflicting groups, as urban manager, and essentially as urban economist (Rydin, 1993).

The type of planner who is involved in this strategic role is primarily a manager rather than a specialist with superior expertise. This may reflect the bureaucratic nature of local government. Those who become chief planning officers, may not be visionaries, reformers, or creative people either, and may not appear to be motivated by any desire to solve urban spatial problems. They have may no policies to offer and none may be required! Rather, they are seen to be good managers, able to organize their staff and resources, and good at public relations and placation. Individuals with personal experience of urban deprivation or with ideas and visions might be looked upon with suspicion as unsettling, but good blokes who are potential team leaders are welcomed. Sue Essex (1991: 86), writing about teaching planning, commented how disappointing and surprising it was for her each year that on asking her new students why they chose planning she found typical answers were that a relative had done a planning course, or that career prospects were good, or that they liked geography at school: not that they wanted to change the world. Jacky Underwood (1991: 147, in the same book, writing on dilemmas in planning practice) comments how this mentality differed from her own desire to fulfil her purpose on the planet, and how her aspirations cut little ice when she entered local government. But the culture of governmental organizations hardly encourages visionaries, revolutionaries or campaigners. See-sawing: at the risk of appearing self-righteous, I echo their sentiments entirely. I felt 'led' to go into town planning, and I still cannot understand how anyone can value it so lightly to see it only as a career.

3

WOMAN IN THE CITY OF MAN

DIVERSITY AND DEFINITIONS

Caution

In this chapter I describe the problem: is the gendered nature of the built environment (cf. RTPI, 1989b: GLC, 1986a). In discussing the city of man, as against the demands of the women and planning movement, one can present a simplistic model of a dualistic situation in which all men are to blame, and all women suffer equally. Therefore within this introductory section I seek to qualify terms such as men, public, and majority which traditionally have excluded women; and to consider differences among women. The situation is complex, as neither men nor women are unitary groups. Each is composed of a range of ages, classes, income groups, and ethnic types, with differing degrees of power. Individuals are affected by the deficiencies in the built environment to varying degrees. There are some major differences of opinion among women as to what ideal town planning policy should be like, for example, on issues such as whether childcare provision should be located in residential or employment areas; on the role of the private motorcar as against public transport; and on emotive everyday life issues like whether dogs should be banned from parks and beaches where children play. Differences of opinion exist among urban feminists as to priorities. Whilst some see children, alongside women, as the last great minority group found to be discriminated against, others believe a pre-occupation with children, and 'homely' issues perpetuates gendered roles. Nevertheless, it is argued, relatively speaking, women *qua* women do share many common problems in the built environment because of their structural position in a patriarchal society, this viewpoint being articulated by urban feminism. The problem is that planners have produced policy on the basis of a gender-blind perspective, to meet the presumed needs of the public (Keeble, 1969). Women often appear to have been excluded from this public, being seen as a separate minority category, existing in the private realm.

Therefore, the chapter kicks off with an explanation of feminism. This is for explanatory purposes for those unfamiliar with the field. Second the issue of class will be reflected upon. This is because class has frequently been seen by planners as the main determinant of differential needs among the population, and thus has been a key determinant of planning policy. Planning has been preoccupied with land-use policy related to work, employment and regional economics. Since much of the discourse had been undertaken in non-gender-specific terms it is important to consider both the applicability of the concept of class to women, and the implications for planning policy of seeing women as part of the workforce. This introductory section is not intended to provide a theoretical base (see chapter 1) but seeks to reflect upon gender and class for sensitizing purposes. I highlight how the social construction of gender and class is constrained by belief in the existence of public/private dichotomies. The way in which class and gender is seen is a major determinant of planning policy (one overstressed, the other underplayed), and thus of city form and structure. I present a codified summary of existing socio-economic classifications and alternative feminist categorizations elsewhere (Greed, 1993a: 53–63).

Gender and feminism

At a structural level feminism provides for many women a meaningful explanations of life, clearly dividing the sacred from the profane (Eliade, 1959), on the basis of gender. (See Appendix II, p.197 for texts.) A broadly feminist perspective may be described as follows. In summary, while it is not contested that women and men are different sexually, in particular because most women can give birth and men cannot, it is held by proponents of feminism that such biological differences should not necessarily determine cultural differences to the disadvantage of women. For example such differences do not have to determine who should be responsible for the childcare and who should go out to work. Such gender roles are seen as being cultural fabrications which are imposed on women because of patriarchy. This is seen as a social system created by men to serve their interests alone to the detriment of women. This does not preclude the possibility that individual men might be more egalitarian in their attitudes to women than others, or that some women might be anti-woman, but structurally, as individuals, men are more likely to benefit from the status quo.

'Man' for the purposes of this research is defined as chiefly white, male and middle-class, but might also include patriarchal women, and also black people who identify with their oppressors ('you can't tell a black man by his colour'), but excludes those men who have sought to take an enlightened perspective towards gender. If I use 'men' it means all men (structurally in society); 'many men' means the generality of men, and 'some men' means individual men who are less representative. Obviously some men have more power than other men, and some men oppress men as well as women. But, relatively speaking, in the public realm of the city all men are in a more powerful position than all women. For example, at the public policy level sporting facilities have been provided as a matter of course broadly for all men. In contrast women planners, along with all other women, may find their demands for better childcare and local, women-related, community facilities ignored. At an individual level, most people live and move in different realms in their daily lives, and gender is not always the primary determinant of how an individual is treated in a particular setting. Class, ideological perspective, and personal persona, as well as gender are important determining factors in how people are treated in any situation. It might be argued that various types of low status men, such as unemployed men, men racially discriminated against, and men in menial occupations, are less powerful than professional women town planners. However, although this might be so in respect of their position in the world of employment, at a personal level, women planners might still be intimidated by some men. A woman might be a chief planning officer but if she walks along the street at night she is just another woman!

Different feminists have different approaches to solving the 'problem'. Liberal feminists might soft-peddle on the concept of patriarchy and believe that men 'don't realize' the effects of their actions because discriminatory attitudes have become almost second nature as culturally normal behaviour (Jagger, 1983). They may think men are planning in the way they do because it seems entirely logical and fair to them on the basis of their male, life experience. Such women are unlikely to see planning as part of a conspiracy against women. They may argue that gender-awareness education for men is needed, backed up by more equal opportunities legislation for women to enable them to compete on the same footing as men. Radical feminists, who see the system of patriarchy as the cause of society's problems, might argue that men realize perfectly well what is going on. They might interpret gendered spatial divisions as deliberately designed to marginalize women, and to keep them in a specific place within society in order to maintain patriarchal power. More radical separatist groups might see men, as a species, as the root of all evil, and demand a complete restructuring of society, based on excluding or limiting men's participation. Socialist feminists might on the other hand argue that whilst the existence of patriarchy is indisputable, it is a mani-festation of capitalism, which is the root cause of inequality, especially class differences (McDowell, 1986).

Reflecting on the concepts of gender and feminism, it is important to stress that many women planners appear to be unconcerned about what sort of feminists they are, or about the niceties of achieving a theoretically correct position, although a person's standpoint undoubtedly affects their strategy for change. Such is the nature of macro-sociological theory concerned with causation that it may be unprovable, however enlightening (Acker, 1984: 36). Many women planners became interested in feminism because they were concerned with physical planning issues which were detrimental to women and which they sought to change. Others became interested in feminist ideas because of their own personal experiences of being, as they see it, discriminated against when they sought promotion, or because of personal experiences of sexual harassment in the workplace or on the street. In conversations with women planners I got the impression that everyone was very busy and few had the time or interest to devote to academic feminist publications and debates. Much of their wisdom came from their own life experiences, through which they developed their own, somewhat eclectic, versions of feminism.

Some women planners appear quite schizophrenic in respect of their public as against private views on urban social policy because of a desire to retain a feminist front, or for fear of not being seen to be politically correct. For example, they may pay for private childcare services since the state provides no alternative to meet their desperate need for domestic help, whilst supporting the concept of public provision for all (*vide* 'Nannygate'). Many a young woman in a planning office has had to endure taunts from her male colleagues on receiving her pay slip each month to the effect, 'you don't need the money; you should be concerned about the working class and the homeless and should give it to them'. It would appear that such advocates of selflessness for others seldom apply the principle to their own lives. They may simply be citing generalized societal problems to diminish the importance of inequalities suffered by women. Indeed many women planners are torn between the need to ascribe to public condemnations of 'money' and 'capitalism' by their left-wing male colleagues, whilst being overwhelmed in their private lives about how they are going to manage financially. Women feel particularly vulnerable should cutbacks occur, because they may be perceived as a reserve labour force. The male colleagues who condemn them for their 'selfishness' are likely to be ensuring their own financial security through local government pension schemes. Women planners are more likely to have discontinuous employment records, and thus are less likely to be benefit from such schemes. The ability to recognize, and survive in, two sets of reality, to think in two ways, and be able to see-saw between the two are essential skills. In conclusion 'gender' constrains women's experience of the urban situation both at the public policy level, and the interpersonal, day-to-day level.

Class and capitalism

I will now reflect upon the problem of class *vis à vis* women and planning, especially the implications of their erstwhile exclusion from the academic discourse of class. Criticism of the planning system has traditionally been framed within urban political analysis (that is within the public realm) in terms of class conflict caused by capitalism, rather than in terms of 'gender' and thus of patriarchy. Structural theories of society, such as that of Marx (1857), have been based on a sexist emphasis on 'male' work and production (or even a 'sexless' image of the worker), that is on work outside the home in the 'public' realm. Little regard was given to women's role in production, reproduction and consumption in the so-called private realm of the home (Walby, 1986). Delamont (1992) in her analysis of the history of sociology, argues that the subject discourse was constructed to prioritize 'work' as epitomized in the interest in (male) social class, whilst aspects of human life which centred around the home and the community were increasingly relegated to the realms of social work as a 'non-academic' women's interest. As Carole Pateman (1992: 227) explains, in relation to Hegel's influence on the way modern man sees the world, paid employment is the key to citizenship, and citizenship is male. Thus, by inference, the working class which the planners say they are planning 'for' really means 'the male population that matters'. Sociology reflects ancient beliefs about the worth of different social groups, and the need to make divisions among them, especially as to the characteristics of the working and middle classes, and who falls which side of the following dichotomies: clean/dirty; mind/body; industrious/lazy. In making such divisions, oddly, the employment of the male head of the household has often been taken as indicative of the likely social, and economic characteristics of the rest of 'his' family. The dichotomies adopted by socialist theorists might be based on inverted criteria, the working classes being seen as possessing the positive attributes in the dichotomies oppressors/oppressed; parasitic/productive; bourgeois/proletariat; pure/impure. Women's concern with childcare and domestic responsibilities does not 'fit' easily within dichotomies, based as they are on a reality comprising an 'adult' world of male productive labour. Women may be seen as parasitic, and feminism as bourgeois and detracting from proletariat interests.

More liberal schools of thought have also left women out. Some women may more readily identify with Weber because he takes into account a range of factors, not just economic issues, in determining a person's status and life chances. When town planning became more sociological (see chapter 9) this created problems for women as the planners took on board the sexist baggage that accompanies the subject. Whilst a social viewpoint often improves the chances of women's issues being considered, it seems to me that, for women, no sociology might be better than sexist sociology (compare Greed, 1991: 116). Women do not fit into the traditional academic discourse, and like ethnic groups, are likely to be treated as an add-on minority category. Tönnies, in defining his two models of society, the old and the new, either *Gemeinschaft* or *Gesellschaft* (corresponding to public urban life or to the private realm of the village community) left women in the awkward position of not quite fitting into either side of the dichotomy (Bernard, 1981: 520). Likewise urban sociological theories as to the differences between pre-industrial and industrial cities do not necessarily work for women (Sjoberg, 1965; and Wirth, 1938).

Women, of course, are workers in their own right, but this is not reflected in official attitudes and socio-economic classifications. The differences amongst men and their occupations have been expressed by a diversity of socio-economic class divisions and theories. The socio-economic groupings defined by the Registrar General in Britain allow for hundreds of obscure gradations of male occupation but little differentiation of female jobs. In contrast, 39 per cent of women fall into the categories of secretarial/clerical and sales staff (Abbott and Wallace, 1990: 33; Arber *et al.*, 1986) and this constitutes the largest occupational group in Britain, numerically greater than the largest male category – skilled manual workers – and one of the largest groupings of commuters travelling to work in the central business district each day. This sameness is a tautological over-simplification, for women in this sector are involved in many different activities at different skill and status levels (compare Crompton and Jones, 1984; Allin and Hunt, 1982). Whilst some men planners might accuse their women counterparts of stereotyping all men as 'the problem', there is a residual tendency for some men planners to perceive all women as non-workers, eternally fixed as housewives or young mothers. (Does that mean the children or the mothers are young? I found many examples of this phrase.) Only 15 per cent of households consist of a male breadwinner, a non-working wife and dependent children (Morphet, 1983; OPCS, 1991: tables 2.3 and 4.9). In some areas over 45 per cent of households consist of one person. This has always been the case in some inner urban areas, a fact disregarded by post-war planners. An increasing proportion of the population are over retirement age, and the majority of these are women, most of whom live longer than men. In parts of the South West, the numbers of elderly women living in single households outnumbers the stereotypical 'young mother' as the 'typical woman'. A large proportion of young mothers are also workers. The details of women's employment and demographic characteristics have been documented by women (Crompton and Sanderson, 1990; Lewis, 1992a: 65–90; McDowell and Pringle, 1992: part III; Walby, 1986) and to a lesser degree by men (Duncan, 1991).

When considering women and planning issues, one has be careful to disaggregate women into different classes. In one of the standard 'geography and gender' texts, *Women in Cities* (Little *et al.*, 1988) the title of the last chapter, 'Working class women, leisure and bingo', suggests that perhaps such women were not entirely included in the rest of the book. Are we all talking about the same women when we sound off about 'women and planning'? Do some middle-class women have as abstract an image of working-class women as their male counterparts? Class differences have a bearing on understanding why different women in planning have varying degrees of career success. Many women of working-class and ethnic minority origin feel deeply betrayed by some middle-class feminists who have ensured the discourse remains one of gender alone without bringing in awkward class differences amongst women. Feminism – and equal opportunties – can be used to legitimate pre-existing class privileges and social demarcations among women. I find it helpful in understanding the differing experiences of individual women to see class as consisting of a composite of factors which make up a person's life chances (Weber, 1964). These include economic class, but also family and social background, and personal characteristics of dress or speech, personal interests, ideological perspective and manners: all heavily gendered attributes. All these factors affect how people are treated, and whether they personally experience class-hatred, or class-love (Webb, 1990: 216). The situation is a complex one, and different for each woman, but relatively speaking, gendered conceptions of class disadvantage all women at the public policy level, and in their daily lives.

CITY FORM AND STRUCTURE

Women in cities

It has been demonstrated by research and human experience that women suffer disadvantage within a built environment that is developed by men, primarily for other men, as explained in published research (as listed in Appendix II). Statistically the average 'man' is a woman, as women constitute 52 per cent of the population (Rogers, 1983; OPCS, 1991: table 1.2) and 80 per cent of women live in urban areas (OPCS, 1991). However they are still seen by many men planners as a minority, and are planned for accordingly. The following account concerns women in cities, but it is also true that rural women experience an additional set of difficulties (Little, 1990; Whatmore, 1991). I base this summary on combined material drawn from a range of reports on women and planning (listed in Appendix II, pp.197–200), and on material from conferences, my travels and conversations in the world of town planning. The editor of *The Planner* kindly published my letter (4 October 1991: 6) inviting planners to write to me giving their experiences and views on the subject. Some women sent me full accounts of their personal experiences which were enlightening and which sensitized me further to the problems women encounter in the built environment. I discuss these problems at three levels: city-wide, district and estate.

City wide planning

Zoning and division

Many women planners consider the nature of macro-level, city-wide form and structure, especially the inter-relationships between land uses and transport systems (foreshadowing D.o.E, 1993), *key* to determining the ease or difficulty with which they can lead their lives. See-sawing: some male planners still underestimate women's ability to deal with such macro-level conceptual issues (see cartoon in *The Planner*, Vol. 79, No.7: 5, July 1993).

Two problems are frequently identified at the city level by women: first the imposition of land-use zoning on urban settlements, and second the outward spread of cities. Both of these create great inconvenience for many women, especially those without cars (WGSG, 1984). Land-use zoning was advocated in the nineteenth century to keep housing away from industrial areas, ostensibly for public health reasons, but arguably also to separate public/private and male/female realms in cities as discussed in later chapters. It has continued into the twentieth century under the pretext of creating greater efficency, convenience and health. Increased flows of commuter traffic, generated by over-zealous zoning, have created new forms of noise, danger and pollution. The division between work and home (which presupposes work happens outside the home) makes it increasingly difficult for women to combine outside work and activities within the home in a flexible manner as distances between zones have grown to motorcar scale. As will be illustrated in later chapters, division has been used as a means of controlling and solving a range of other urban problems, reflecting deeper beliefs in the importance of dichotomization as a means of creating a manageable reality.

Not only are work and home divided, but different types of women workers are divided from each other along class lines in different residential areas. In British cities zoning is restricted to land uses, but in North America zoning can be based on density and housetype too, which creates a socially divisive pattern. Control of social aspects as against land-use matters is a double-edged sword for women. In seeking to provide low-cost housing for rent by making housing tenure a legally 'material' planning issue, as stated in Circular 7/91 and updated PPG 3 'Housing' (D.o.E., 1992e), a precedent may have been established which could be used to increase patriarchal control in residential areas. But the extent and zoning of suburbs in Britain is nothing like that found in the 'burbs' of North America or Australasia. There is a danger in citing classic American accounts of the suburbs such as Gans' *Levittowners* (1967) because one can import inappropriate foreign images of suburban life. Typically, these housewives are portrayed as spending much of their time having coffee mornings, enforcing stereotypes of women as lazy. However, early feminist critiques of suburbia such as Friedan (1963) seemed (to me at least) to be to do with middle-class, American housewives who did not work outside the home. Whilst condemning negative stereotypes, one must acknowledge cultural differences. North America is generally more affluent than Britain, houses are larger, and cities are much more dispersed, with suburbs extending over 50 miles from the centre.

Outward growth and transportation

Planning has been preoccupied with the need to control the city, by trying either to contain it, or to decentralize it. Linked to the latter aim were ideas of thinning cities out, presumably to neutralize the evils of the slums. The ideal density was calculated with a view to create 'balance' (a popular planning word which can justify just about anything). Paradoxically, speculative housing development by private developers has been looked upon as sprawl – in contrast to the planned decentralization of public (social housing) schemes, although both may be seen as eating into the countryside. The way in which planners judge the merits of a scheme is influenced by the question of whether it is a public or private development. This reflects ancient beliefs about the profanity of trade, and by association the evils of private development, as against the merit of public service and thus of social housing. For women living out on the edges of cities without cars, and limited public transport, the question of whether their estate is a private development or a public council scheme is academic as the problem is similar in both types of development.

Eighty per cent of households own at least one car and 21 per cent own two or more (OPCS, 1991: Chart 9.4: 150; Greed, 1993a: 12), but it does not follow that women have equal access to the car in the daytime, as it may be taken away for the day by the breadwinner. Forty per cent of females, as against 80 per cent of men, have a driving licence, but only 15 per cent of women have the use of a car in the daytime and 75 per cent of car journeys are made by men (Pickup, 1984; RTPI, 1991; RTPI, SW Branch, 1991). But, 75 per cent of all journeys are not by car, as most women do a lot of walking (compare GLC, 1985). The train-spotter mentality of some male planners, who only see journeys made by mechanical vehicles as worth counting, does not bode well for women and walkers. The emphasis in policy land use–transportation planning has been upon accommodating the journey to work (traditionally perceived to be taken by the male breadwinner), which is seen as mono-dimensional: home to work and back. Women, on the other hand, might be combining home → childminder → school → work → shops → home, and travel at off-peak times of the day or in other configurations, if they are part-time or shift workers (i.e. they are moving between the public and private realms). Women shop assistants and office workers may not be perceived as commuters, because the male journey to work in central areas dominates, and renders female journeys invisible. There used to be local planning authorities that only asked the head of the household about his journeys when under-taking traffic surveys. Women are often perceived as static, existing solely within the residential area. Many of women's problems with planning relate to their daily activities and movements among land uses, that is, how they 'uze' the different uses (p.22). Some women can only resolve this problem by using a car, for which they may be further judged, even when there is no alternative.

The problem is compounded by inadequate public transport, which was once a wonderful system before mass use of the private motorcar became commonplace, especially among men. But men are not a unitary group, and it was in the interests of powerful men in the motorcar industry to neutralize public transport. In the United States this was done by the large car-manufacturing companies buying up and closing down pre-existing public transit systems (Wajcman, 1991: 128). In Britain in the 1960s it was achieved by the Beeching railway cuts, which reduced the number of railway stations from 5,000 in 1958 to 2,500 in 1968, closing many branch lines just at the time when expanding commuter villages needed them (Greed, 1993a: 66). Some might argue this is not a gender issue but one of car ownership and of rampant capitalism. In fact, men's control of women's mobility is a centuries-old phenomenon, also reflected in fashions such as footbinding (Rossi, 1977), hobble skirts, and high-heeled shoes, and reinforced by urban design. In the United States poor women's mobility is constrained by the requirement that they must cash their food welfare stamps only at certain food stores. North American cities are weak on the provision of cross-city public transport, with everything going into the centre and little chance of potentially shorter, cheaper tangential journeys between residential districts.

Women's lack of mobility is often seen as their own fault and not an urban structural issue. Women's justified fear for their safety, an issue linked to travel, may be diagnosed as a disease (Ussher, 1991) such as agoraphobia, or even schizophrenia, 'an inability to distinguish reality from unreality'. See-sawing: I remember as a student being told by a male student, when commenting on the problems of transport for women, 'well that's their fault: they should learn to drive'. Such attitudes persist in modern textbooks. Focas (1989: 168) states that a national policy of encouraging women to take driving lessons, and to use cars available to them would alleviate the problems of constrained accessibility suffered by many women who are non-drivers in car-owning households. People may not be allowed mobility if they want public transport instead, especially in the case of keeping ethnic minority groups out of certain areas. Wajcman (1991: 133) explains how 200 low hanging over-passes on Long Island were designed specifically (between the 1920s and 1970s) to keep buses and goods vehicles off the new parkways, thus keeping the poor and black people, who in North America are the main users of public transport, away from the recreational areas to which these routes led. Age as well as gender restrictions on mobility can result from, for example, reducing the numbers of public lavatories (which affects children and the elderly as well as women). Men have approximately three times the number of facilities than women (WDS, 1990). The reduction in the number of public conveniences is a matter of national concern, not solved by the substitution of the much feared and hated, impractical and unecological 'Superloo'. This cause is championed by Susan Cunningham (Cunningham and Nortan, 1993) through the 'All Mod Cons' campaign (Tuck, 1993). Poor street lighting, plus taking off late-night buses, may also affect young teenagers' mobility, whilst adding steps and changes of level to a new development discourages pushchair and wheelchair access and hinders the progress of the elderly.

Nowadays the wheel has turned full circle so that it is politically correct, and environmentally friendly, to be anti-car and pro-road-pricing to reduce the numbers of vehicles. But many women, especially those who live in the suburbs, cannot put the clock back. They and their families have located in the suburbs in good faith because in the past the town planning system endorsed decentralization. It seems to me that a small number of middle-class people who, having recolonized and gentrified parts of the inner city so that they can now walk to work, appear most unwilling to let others park in central areas, even when those others are the very people who work for them in their central offices. Some women are wary of park and ride schemes for fear of being stranded by unreliable shuttle buses. Others dread returning to their car in the evening to find their car has been clamped. In both cases women fear such measures render them more vulnerable to urban crime. Many planners report that members of the general public are highly suspicious of road-pricing policies, because they imagine that their aim is to get everyone else off the road so that those with power (including the planners) can return to the traffic-free road situation which existed 30 years ago (as discussed in RTPI South West Branch, 1991). Those who cannot pay (particularly women) will be encouraged to use public transport, or lose their jobs if there is none, regardless of how unfair this is (cf: Rogers, 1983: chapter 6: 'Have you got a company car?').

Indeed a new species of macho-green man has been spotted in the world of transportation who is even worse than the traditional suit-clad, car-borne male transportation planner of yesteryear. In conversations with women planners, and from personal observation, it would seem that some young men on bicycles (dressed in lycra outfits, face covered with air-filter masks) are extremely arrogant and aggressive, just like some men in cars, and seldom are they burdened with shopping or children. Changing from cars to bicycles might be greener but is not necessarily less sexist. One hears of incidences of young children being mowed down on shared footpaths/cycleways (a design which in itself reflects the low importance given to pedestrians' rights), particularly from sources in Milton Keynes. Alternative forms and patterns of transport will enforce the problem for women if they are still based on sexist assumptions about women's journeys, however green the intentions. Proposals for restrictions of car use to essential users may exclude many women because of likely relegation of women's non-work journeys (to shops, schools, doctors) to the category of leisure trips.

District planning

This is the intermediate level between city-wide policy and a consideration of local estate design principles, at which level the key question is how the local area is structured internally and inter-relates to the city as a whole. At this level, zoning problems combined with the capriciously dispersed location of essential local facilities such as schools, shops, and health facilities has compounded the problem for women, creating time-budgeting problems and sheer inconvenience (Bhride, 1987). This affects women of all classes, whether they be factory workers struggling between decentralized industrial estates and inner-city schools, or office workers who have to decide whether to risk having the time to do the shopping in the evening before the shops close or attempt to do their shopping in the lunch hour (Bowlby, 1988, 1989). See-sawing: if they choose the latter they are likely to find diminishing numbers of food shops in the town centre, and have to endure the comments of their male colleagues, who may see it as unprofessional to have a pile of shopping bags hidden under the desk because this is an intrusion of artefacts from the private realm into public world. Over 60 per cent of the occupants of central offices are women. They are thus the largest sector of commuters and workers in this area (Greed, 1991: 173), but neither land-use patterns nor local facilities distribution are designed with their needs in mind. Women with cars may be at an advantage in getting between land uses but car-parking problems have become so intense that car ownership is not always an advantage.

In the case of local shopping provision, much of the problem revolves around how 'shopping' is seen. If a male planner looks at a shopping centre proposal it may trigger quite different reactions compared with his female colleague as to what the planning considerations are. The former is more likely to consider externalities such as traffic generation and car parking, whilst the latter is more likely to consider internalities in terms of actually 'uzing' the scheme (a point raised by my part-time mature students). Indeed, 'going shopping' becomes transmuted into a male activity called, 'retail development' in the property world, dominated by impersonal retail gravity models (Roberts, 1974) and financial investment factors. (This is so over-whelming, that at one stage I felt inadequate to teach on the subject of retail development.) Shopping is such an unfamiliar activity for some men planners that they have to go out of the office to find out what any woman could tell them from daily experience. Once, when I was out shopping, I was suprised to find a male planning colleague in the supermarket, undertaking a spot survey of the price of goods. He assured me that this information would be used to help determine the future retail planning policy. No doubt in his mind this was a reliable survey, because the data was quantifiable, being used alongside information on provision of parking provision and retail floorspace (Roberts, 1974). But to me quantitative factors seemed relatively unimportant (and changeable) in comparision with the many qualitative factors which determine the 'attractiveness' of a shopping centre, such as the convenience of its design, its accessibility by public transport, the provision of public conveniences, and whether the shops are open outside office hours. My encounter highlighted the fact that men have to cross the public/private and work/home divide to gather information 'in the field' – and for them it can be a terrifying journey into unknown territory – to plan for facilities they neither use nor understand.

In planning for existing local areas, many planners did not appear to understand the value of organic development which has grown up over the years, in a manner which creates a mixture of land uses and facilities convenient to the residents. This arrangement might simply be seen as acres of chaos by the planners. When the planners designed their New Towns, districts were structured on what they saw as rational, scientific principles using land-use zoning, but the end result was far less suitable than unplanned development. At the local level, they sought to create balanced neighbourhood units, in which community spirit was meant to blossom. (In contrast, inner-city 'hoods' are looked upon with condemnation as deviant areas.) Designing New Towns on the basis of the neighbourhood unit might be seen as another manifestation of the fondness for division in planning policy, creating distinct 'women and children' zones separate from the real city. Ironically although neighbourhood units might be seen as falling on the domestic, female, homely side of the public/private dichotomy they may be most inconvenient for women to inhabit, often with shops, schools and factories in opposite directions, involving additional walking and worry (Attfield, 1989). The location of factories was often done with no regard to the fact that indus-trialists saw the availability of cheap part-time female labour as a key factor in relocation. In seeking to create an image of newness, purity and cleanliness, dirty essential uses were often given low priority. A certain New Town was built without space for a rubbish tip or for scrap-metal dealers. Another only provided a maternity hospital in the last phase of the development after most residents had completed their families.

At the local level, there was nothing to keep the neighbourhood unit fixed; it was an artificial concept to begin with, based on a sentimentalized ideal of community. 'Community' is a fascinating word wheeled out when the planning of the working class, ethnic minorities, women, single-parent families, and other 'problems' are under consideration: a zone perceived as marginal to the public realm of the real world of the male majority. As one woman planner wrote to me, giving her views of a real New Town where she worked which was famed for being designed originally to ensure a sense of community, 'I would describe Blank New Town as a video/takeaway in front of the TV [town]'. Indeed the planners' obsession with creating a sense of community whether it be in New Towns or inner cities, has been criticized as banal and insulting by many, especially members of ethnic minority groups. As Eno Amooquaye writes, 'words like community are bandied around without honesty of belief in its existence' (letter in *Planning* 10 April 1992: 2; cf. Griffith and Amooquaye, 1989; Krishnayaran, 1990). Higher density planned solutions with centralized facilities as found in parts of Europe are not necessarily any better, because although quite different in layout, they are still designed around patriarchal values (Strauch and Wirthwein, 1989; EC, 1990; Van Vliet, 1988). Many French women, for example, prefer living in traditional central arrondissements 'with the sounds of active Europe' all around them (Cardinal, 1991: 6) to dwelling in the silence of Anglo-American suburbs (not that many women French or English actually have the chance to compare the alternatives). National differences have to be acknowledged as there may be no one urban solution ideal for all women. It is extremes of density, high or low, combined with poor design standards, which are the problem for women. Many British (and North American) women would say that they love their gardens and need them for the children, dogs and cats, and are unhappy with the higher densities and more sophisticated European urban lifestyles. The problem may be not the suburban style of housing itself, which is generally liked, but rather the lack of integration of other supporting land uses and amenities in suburban areas, and their remote location relative to the centre of the city.

Estate planning

The estate layout level is the habitat of another sub-subcultural world within the town planning profession, a world of housing layouts and estate design, almost totally dominated by male architect-planners. This is a realm of town planning practice where, potentially, great damage can be done. Unenlightened policies can be imposed on women's lives at the most detailed level of planning control and design. Well-intentioned policies may prove damaging for women if they are based on sentimental and unrealistic stereotypes of the domestic realm, or on gendered perceptions of the nature, activities and roles of women and of children. Portraying children primarily as male, as lovable urchins (cf. Ward, 1978a) may detract from the need for female children too to have play space and freedom within the built environment. Some male planners may actively seek to resist women-friendly design, which is seen as limiting and likely to mess up their plans. In the same way designing for the disabled is often seen as being visually intrusive (so much for the architectural adage 'form follows function'), as discussed in the tellingly entitled article, 'Disabled? don't spoil our design' (LBDRT, 1991: 6–8). Women worry because, for example, they have to deal with *those* steps outside their front door. But one must be careful in saying so, because one might be told this just shows how small women's horizons are, and how 'helpless' they are. However, one cannot, necessarily blame the planners for all bad design as in many cases they have limited control over the standardized designs produced by large-scale, speculative, suburban housing estate developers.

Town planning is chiefly concerned with the external layout and design of housing estates. Planners have limited control over the internal design and space standards of houses, or over the size, numbers or types of rooms. This may be accounted for by the fact that internal design is seen as the province of the architect, not the planner. Many women planners consider that it is unrealistic for planners to deal with the outside but ignore the inside of housing. However, in the past planners had more control on the internal design at least of public (social) housing (Parker Morris, 1961), but limited control over private, owner-occupied housing. The building regulations, however, control construction standards on all buildings, but require very minimum space standards for residential rooms. With the very small size of some houses, and the problems of lack of external child play space, lack of internal space and thus of personal privacy can be a major problem (compare Musgrove, 1992, which describes the work of Jean Hillier on this matter). (Jean Hillier incidentally, left Britain to become a professor of town planning in Perth, Australia.) The woman of the house is unlikely to have a room of her own (Woolf, 1929). But, many women would be most unhappy about greater state control over the private sector, unless women dictated the standards, as it might restrict their freedom within an already imperfect situation rather than providing alternative provision. Under the Children Act 1989, space standards are being applied to childcare provision (even when it is only for a few hours with relatives): a hindrance in view of the lack of provision and unwillingness of some local authorities to register new childminders. This public/private, outsides/insides debate, as to the extent of planning control over housing design, is a theme which I will develop further in the historical chapters.

At the detailed level of the estate outside the dwelling, problems revolve around the issues of pedestrian access, steps, lighting, surveillance, planting, and play areas. Women, as the majority of pedestrians, are critical of the little details like the use of rugged paving stones which are meant to enliven the streetscape. These can shake the wheels off a buggy, or twist people's ankles, and wrongly aligned drain covers can catch bicycle and pushchair wheels. Likewise, changes in surface colour and texture are meant to create an interesting townscape and soften the division between vehicular and pedestrian zones on housing estates. As a result pedestrians have to watch out for cars trespassing onto pedestrian areas, and seek to avoid walking in puddles of motor oil left from weekend repairs on shared surfaces. Women have criticized the principles of housing layout expressed in the Essex Design Guide (1973), master-minded by the architect Tony Aspinal, which was so warmly welcomed by fellow planners. Women dislike exciting elements which create suprise and interest in the urban environment, such as blind corners, and meandering indirect paths. High walls alongside footpaths, intended to create a sense of urbanity and to ensure garden privacy, are particularly disliked (MATRIX, 1984: 50; Collier, 1991).

In considering detailed design issues, it is important to distinguish between the fact that some problems affect women with children, and some affect women without accompanying children. There is a tendency to conflate the needs of women and children so that play areas are often presented as a woman's issue as if the mother herself is going to play on the swings. Other women working from home, and people on night shifts, do not want such facilities integrated into the residential layout if it results in (other people's) 'yelling kids' playing football beside their house. Houses in some New Town schemes are marooned in swathes of meaningless landscaped public space making everything unnecessarily spread out and windswept. This causes anger if individual private plots are very small. This over-enthusiasm for grass and open space was identified by Jane Jacobs (1964) as a characteristic of Anglo-American planning, but one which was particularly prevalent in British New Towns. Many working women perceive landscaping consisting of vast useless areas of grass around out-of-town business parks as just as much a nuisance because it is a means of preventing the growth of all the jumble of shops and facilities which makes an area liveable in.

MAKING INTERCONNECTIONS

When I read through the various sources containing women and planning policy statements I am struck by how much of the material is not about spatial land use and development, but rather concerns aspatial (social) considerations (Foley, 1964: 37). This reflects the realistic fears women have about their personal safety in urban areas. Whilst many of the observations relate to new developments, in Britain new-build represents less than 3 per cent of the total building stock in any year, a good deal less in the current property recession (JFCCI, 1991). Many problems are created for women when no thought is given to the implications for women of policy decisions because it is presumed the topic in question has nothing to do with gender. For example, as a result of a pedestrianization scheme in a certain provincial town centre, buses were redirected along a back street. The bus stops were relocated beside a somewhat desolate car park with some unsalubrious public lavatories just across the road. Now on dark evenings hardly anyone uses these bus stops; presumably they walk a considerable distance back to the terminus or on to the next stop.

The problem is often presented as an urban issue, but women can experience similar fears in the countryside, especially when there is no lighting at all. Recently a West Country village refused the offer of street lighting by the council because it was seen as detracting from the historical authenticity of the rustic scene by the townies who had bought up houses and gentrified the area, although many women and non-car drivers in the village wanted it installed. Also women who have to travel around in both urban and rural areas in their work (such as surveyors and planners) soon find that public facilities in an area are not for everyone. There might be a vast number of public houses in an area, but many women would not be willing to go into them on their own. Women are further restricted if they are travelling with under-age children, including babies (Hey, 1986) which are not permitted in pubs. Yet there may be no other buildings offering lavatories, or refreshments in the vicinity. The government has (March 1993) made draft proposals to allow children to accompany adults into pubs, as they see a family atmosphere having a civilizing influence. This is in order to reduce drink-related crime rates – not for the benefit of women and children. At a women and planning conference I attended a discussion group in which one of the problems women encountered whilst travelling with children in the evenings was raised: namely the difficulty of finding any public conveniences open. The only man in the group (who until then was giving a convincing impression of being a new man) cheerfully commented 'well they can always go into a club'.

Similar observations can be made about women's problems with sport and recreational facilities. There may be no crèches, proper changing facilities or women-only sessions, and women may experience a lot of hassle from staff and harassment from young sportsmen, ranging from sexist comments to dive-bombing of female swimmers in mixed pools. But if a woman is 'with' a man, without children, and/or over a certain age, the whole situation can change, and she may freely use pubs, sports facilities and rural leisure facilities. Women town planners may encounter problems out on site on their own, whether they be students doing projects or practitioners doing surveys, (Lamplugh, 1989; Woods and Whitehead, 1993). Sometimes it is thought that they are soliciting, as in the case of two women chartered surveyors who, when they were making site visits with their male professional counterparts (Greed, 1991: 156), were arrested by the police because they were seen going into derelict buildings with men, reflecting deeper assumptions and ancient beliefs about women's role and place in the city.

To sum up, the reader should keep in mind the themes of division and zoning (of public/private; insides/outsides), density control, decentralization, balance, and the importance given to the public open space, as characteristics of modern planning, as in subsequent chapters I will seek to unravel 'how', 'when', and 'why' they became important within the professional litany.

4

PLANNING
The spirit of the age

WHAT'S BEHIND PLANNING?

Definitions and generalizations

In this chapter I step back from town planning and explore the gendered nature of planning. It is a key ideology, dominant philosophy and way of doing things in the twentieth century, the *zeitgeist*, the spirit of the age (Casson, 1978). I am not seeking to prove again that patriarchy exists (q.v. Appendix II, p.197) but seek to explore its influence on women as instrumentalized through planning. In particular, I highlight how the planning mentality manifests dichotomized thinking patterns, dividing public from private realms to the disadvantage of women and other minority groups. I will illustrate this tendency with reference to socio-economic, population, third-world, green, and *beaux-artes* planning. I include the latter to show that characteristics which might be attributed to a modern scientific mentality are also associated with historical forms of planning, which manifest the 'same' patriarchal desire for order through division.

As to definitions, anybody can call themselves a planner, and plan (Hall, 1989: 10–17). Individuals who have a wide, governmental, strategic policy-making role may call themselves planners, as may those who have a narrower, professional-body affiliation and possess a specialist, substantive body of expertise (such as town planners). I see the planner, for the purposes of this chapter, as a high-level policy maker (the philosopher king planner, p.30) who is likely to be working for a national or global agency. To have reached such a position the planner is likely to manifest supremely the most valued attributes of patriarchy, not least a scientific and objective approach to policy-making. Science may be defined as knowledge ascertained by observation and experiment which has been critically tested and proved – according to malestream dictionaries that is. The planner as urban scientist is likely to be a man, and more specifically a white, western, highly educated, middle-class one (see p.35 for definitions of 'man'). However, a few planners might be non-western, female, and/or not white. But at the highest levels of planning, professional orientation and belief are likely to override the effects of gender and ethnicity. In this chapter I purposely interpose my own private and opinionated views about the public global forms of planning under discussion, precisely because such planning often appears impersonal, clinical and people-less. I am aware that other women planners might not agree with my views. Some colleagues are of the opinion that the mentality described was stronger in the past in town planning, especially in the 1960s. Others hold that it is as powerful today, but manifest in other types of planning.

Nature and science

It seems to me that planners have a profound distrust of natural processes and of other human beings' ability to act reasonably. Their policies are informed by seeing the world through the lens of science/nature, order/chaos, and mind/body dichotomies (p.12). Man's ideas elevated into 'science' are seen as superior to the processes of nature, chance and destiny, the commonsense of the masses, or the intuitive powers of women. Planning as an ideology embodies deep-seated preconceptions emanating from the Enlightenment (Hampson, 1968) as to the centrality of science in a mechanistic universe. Science manifests patriarchal values *par excellence* (Kirkup and Keller, 1992: chapter 1). I find the inter-relationship between science and planning another teaser. Although some planners like to describe planning as a science, the nature of the beliefs which inform planning suggest one is dealing with a less than scientific profession. In the same way that the surveying profession created a technological smokescreen to cover its true activities (Greed, 1991: 100), planning has often projected a spurious scientific image, possibly to increase its status as a profession and divert criticism from its policies. A scientistic approach is influential in other types of planning, such as economic development, population policies, and third-world planning (as expounded by Rogers, 1980). Such planning is characterized by large scale, future orientated, impersonal, abstract, quantitative approaches to problem solving, whose emphasis on objectivity excludes women's needs from the terms of reference.

It would seem that man is desperate to be different and to separate himself from nature, a category into which woman has been consigned as part of the 'other' (de Beauvoir, 1974: 51) over which man must have dominion as centre of the universe. McDowell and Pringle (1992: Part I) discuss this otherness wonderfully, linking it to how the public and private realms are defined. The image of man dominating nature has also been manifested in western civilization in the celebration of the explorer, conqueror and colonizer of the unknown 'other', be it Darkest Africa, Frontierland, or the female psyche (Haraway, 1989; Freud, 1935). Modern man still imagines himself going where no man has gone before into outer space: the last frontier and ultimate 'other' which must be conquered. The desire to rise above the temporal realm is ancient. It has been expressed, variously, in the Aristotlean desire for man to achieve perfection by the separation of mind from body, and in the ambition of the Christian gnostics to transcend matter and attain purity through perfect knowledge (Coole, 1988). The planner must discover, conquer, and control the problem, which is typically seen as being caused either by 'mother nature' as manifested in environmental problems and over-population, or by human nature. The planner sees intervention in an imperfect world as essential to save society, the globe or the city from disaster. Women are often blamed by planners for creating the problem in the first place: for being feckless, and having too many children, presumably because man fell to their temptation, because women were not adequately controlled. They have been portrayed by the fathers of the church as bereft of the ability to make sensible choices and decisions (Holloway, 1991), which suggests that common patriarchal beliefs about women underlie aspects of both religion and science. Because modern man takes a detached 'scientific' view from 'above', he may see himself as separate from the problem. See-sawing: I feel that this 'impersonal' stance affects the planner's sense of (or lack of) personal accountability. Even when man the planner adopts an equal opportunities policy it is purely a public, amoral matter. He is detached, having no sense of guilt or personal blame, and feeling no need for forgiveness from women or God, or to make restitution and undergo repentence (Holloway, 1991).

Creating realities

It seems to me that planning is not really about saving society from disasters or problem-solving, it is about creating realities, which, as Eliade identified (1959), is one of the functions of a religion, thus blotting out other realities from the terms of reference/relevance. The problems which legitimate planners' intervention in society are carefully chosen (even custom-built) and only selective aspects of a problem are highlighted according to class and gender priorities. Acceptance of the solution which is to be planned for is predicated upon believing in a certain world view, which is culturally constructed and by no means neutral. The process of defining the terms of reference, and creating realities as to how the problem and the solution are seen, often renders women's existence irrelevant, and therefore cloaks over their 'real' needs and problems; the female world is blotted out by the world created by the planners. The key to plan convincingly may be to take what is nothing more than a belief – a piece of imagination – and with immense self-confidence turn it into something factual, objective and technical: the only scientific logical solution towards attaining a reasonable allocation of scarce urban resources for the good of society. Computers, statistics and plans are useful tools in the process of legitimating one's world view. For example in town planning, by the use of these tools transportation (introduced in chapter 3) has been socially constructed to centre on the problems of car drivers, at the expense of public-transport users who are chiefly women. Indeed, the marginalization of women's needs is not special to the world of town planning, but reinforced by the way in which other government departments present statistics, often rendering women invisible in the process (Allin and Hunt, 1982). One needs a readily cowed society to get away with this sort of thing, and it amazes me more women do not complain. We are all taught (especially by teachers at all-girls' schools, Evans, 1991) to 'trust' our betters in the Welfare State, and not to be 'selfish' (women's greatest cause of guilt, Gilligan, 1982), but to care for 'others'. But, many feminists would argue that men planners do not see the people as 'others', but as 'the other' which they as superior beings have control over, and are at pains to distance themselves from: a theme strong in Haraway's writings (1991). Women planners, as both planner and planned, have to be man and other at the same time (McDowell, 1983).

Because planning is concerned with policy-making, it is focused on the future, often at the expense of dealing with present-day problems. The classic excuse for inactivity on equal rights, for example, is that it takes 'time' and change can't be rushed, and we've all got to be patient a little longer. It seems to me, particularly in respect of the content of planning education and textbooks, that planners spend a great deal of time and effort pointing out how bad everything was in the past in order to legitimate their right to plan, now, for the future. For example, the impression is given that in the nineteenth century *everyone* was living in slums and racked by cholera. The emphasis on future planning, combined with frequent allusions to the past, narrows the space available within the planning discourse for discussion of present-day practical problems; so the present becomes shorter. Women are expected to continue to put up with the ostensibly 'temporary' inconveniences of the present situation, so that planners can devote their 'valuable' time to planning for the greater good in the future. Policies related to the future are difficult to argue with because of the long-term time scale and the inevitable element of unreality and abstraction that the topic creates. Indeed, planners may feel restricted in their brainstorming efforts to create future scenarios if reminded of present factual realities. The title of Susskind's article (1984), 'I'd rather invent the future than discover it' says it all. One role of the philosopher king, in 'his' capacity as town planning theorist and seer is to 'create' alternative futures, and divide the future from present and past.

Although planning has often been portrayed as a science, to me it seems much more like a religion, or an opium for the elite. This is because one clearly has to have faith that a future heaven is ahead. The emphasis on the future, and thus the deferred satisfaction, both in terms of the 'planned' having to wait and the 'planners' anticipating a time when they will be able to implement their policies without hindrance, means that planners may live in a dream world. They may pretend that the real world around them is irrelevant, or only a temporary phase before they establish the new millenium. This scenario is redolent with parallels to the other-worldly mentality found in many radical political and religious groups who wait until 'after the revolution' or 'until the Lord returns', or amongst traditional, Cinderella women who wait for Prince Charming. This is damaging for women who, because of their biological time clocks ticking away, want the Kingdom to come now on Earth, not just later in Heaven. Eric Reade comments on the need for planners to have faith in the ideology of planning and ignore the gaps, contradictions and unanswered questions (1987: 98–114). There has always been a utopian thread running through the development of town planning that has reflected both a desire to create new realities, and to escape existing realities. One has to be cautious as to how one interprets planning literature for, in spite of its hard scientific image, much of it might be about unachievable reality.

OTHER MANIFESTATIONS OF PLANNING

Does space matter?

Town planning is mainly a spatial form of planning, whereas the other types to be discussed (except for *beaux-arts* planning) are aspatial, being concerned with social, economic and political issues. None the less they affect the nature of the built environment, albeit by a circuitous root. In my previous book (Greed, 1991) I made much of the importance of spatial policies as a means of maintaining patriarchal control. For to be male is to occupy – and control–space (Cockburn, 1985: 213). Now that feminists have rumbled the importance of space in maintaining patriarchy, attempts are being made by patriarchy through academia to negate its importance. From conversations with friends engaged in sociological research I have found that similar mechanisms are at work in other academic realms, negating the importance of other key conceptual components in the feminist argument. No sooner had feminist scholars begun to challenge the nature of social divisions, showing that gender was a primary consideration and that existing conceptualizations of class were unworkable in respect of women (Abbott and Wallace, 1990), than post-modernists and post-structuralists began to suggest that 'class' was never really very important as a concept after all, and that, well, gender was only one concept amongst many.

A spatial expertise is 'that exclusive piece of specialist knowledge' that the profession possesses to justify its existence and exercise of power (Saks, 1983). Nevertheless space *is* important. We carry out our lives with the assistance of buildings, roads, sewers and drains; we are corporeal beings and, as Harvey (1975: 24) stated, we do not live in a spaceless world. As stated (chapter 2), legally, the town planning system is required to concern itself chiefly with spatial land use and development issues. Although this spatial emphasis can be used to marginalize women's needs as not being land-use matters – as *ultra vires* – women do want to retain a spatial emphasis, on their own terms. This is because so many of the matters which affect women's use of the city are practical, spatial matters as discussed in chapter 2. From talking to fellow women planners, I am aware that there is a fear among them that, in moving away from the spatial emphasis of planning to 'airy-fairy' theoretical debates, attention will be diverted from the specific, 'nitty-gritty' physical planning demands made by the women and planning movement. Of course, a concern with spatial issues within the arena of town planning does not preclude a concern with wider social, economic, and political forces, but they might be better tackled through other agencies or strategies. I am focusing on the effects of gender on the built environment, but this is only one small part of the total problem, because gendered perceptions shape all aspects of human life. The value of adopting a spatial approach to problem-solving has always been open to question. For example, town planning went through an anti-space phase in the 1970s under the influence of neo-Marxian urban sociology (cf. Castells, 1977). Town planning has frequently oscillated between physical and social agendas. It has also been strongly involved in economic policy-making, particularly at the regional level and more recently with ecological issues at European and global levels. The history of town planning has been not so much about the containment of urban England (Hall, 1977a) as the containment of the sprawl of the profession, which various interest groups have sought to use for wider goals, often seeking to undertake social or economic planning under the cloak of town planning.

Social and economic planning

There is always a temptation to move from purely physical land-use planning to the next 'logical' stage, and seek to control the economic and social processes which create the demand for the land uses in the first place. There needs to be some convincing political or economic legitimation for this move, to enable the planner to take on the role of technocrat and ruler. Town planners found socialism an ideal ally in the 1940s and 1960s, because they could argue they were planning for the benefit of the working class: a group which was not usually seen to include women. The need for planning may also be invoked in the name of efficiency and liberal equality within a mixed economy. In this case planners act as powerful urban managers (Pahl, 1977) to enable the allocation of scarce resources in the most rational and profitable manner, in order to increase 'productivity' by the private sector (Ashworth, 1968: 2). The problem with this for women is that many of their needs are non-profit-making and are therefore marginalized as merely 'social'.

In many erstwhile socialist societies, the needs of women were unlikely to be considered as important if production, particularly male factory work, was defined as real work. Women's traditional work might be seen as merely consumption or part of the private cultural, folkloric superstructure over and above the public modern economic base (Buckley, 1989). Exclusion of women's issues can also be observed when socialist states revert to capitalism (Duchen, 1992a), as the primary agenda remains one of 'public' economics. Women in a third-world country in the process of modernizing or developing may find that land, previously communally held, is given to the men. New businesses (and related credit and support) will be put in men's names, even when the women are traditionally the main traders (Rogers, 1980). Likewise, in Britain, women may have great difficulty getting credit to set up businesses. It is often said that women are no good at management and are a bad investment risk. Women's 'other' work is not counted in the Gross National Product, or as collateral for a loan. The delegation of much of women's production to the realms of consumption ensures that their work is seen as charity or duty, not employment. Many women identify readily with the right, for at least capitalism acknowledges their needs provided there is a profit in it (Greed, 1991: 22).

Self-employment can be seen as a means of escape from the system, an 'unplannable', deviant activity, or a form of enpowerment for powerless people. Far from being a form of incipient capitalism, it can be a means of sheer survival for outsiders who, because of discrimination, cannot obtain waged employment and thus become members of the respectable proletariat. Setting up a business is a classic route out of the ghetto for minorities and underclasses (Greed, 1991: 54; Hertz, 1986). Generally speaking, socialism has tended to lump together the activities of large, capitalist male enterprises, and small businesses run by women or low-status men, as being equally wicked, without acknowledging their differing power levels (cf. Saunders, 1979). Many women consider the right to own land or money an equal-rights issue, and feel betrayed when they are told this is selfish or counter-revolutionary. Women often bear the brunt of the work in family businesses, suffering emotional stress and sheer exhaustion. They also comprise the low-paid workforce doing the unskilled, dirty and heavy toil to support the noble socialist worker in state industry. Women are now expected to support the return to family farming in the CIS (former Soviet Union) with their labour, but are unlikely to share the ideals of inheritance and wealth and of being 'master of one's own destiny' currently being promoted (Corrin, 1993). Public discussions of 'market economies' and 'employment' blot out such private realities.

Trade has an ambiguous position in the dichotomies. It can be seen as 'male' or 'female', changing its status according to which 'side' it is allocated. Religion and agriculture also shift sides over the centuries, as do town and country. In former socialist states the official reason for opposition to women acting as small traders was to prevent incipient capitalism. Another reason might be fear of such women acting as zone zappers and moving into the public realm. Men feel threatened by them, therefore they have to be controlled (Buckley, 1989: 110). In pre-revolutionary Russia there was a tradition of women, as small traders, match-makers and healers, 'doing business': a dichotomy-defying combination of spiritual and entrepreneurial activity (Clements *et al.*, 1991) reflecting the economy/culture debate. This was often their only means of subsistence. Similar problems exist in North Vietnam. Brazier states in the *New Internationalist* (1991: 5) that the theory behind the ban on small peasant businesses, mainly run by women, was that distributing goods by the market was seen as inefficient and unjust. But, bureaucrats in distant offices were unable to anticipate what goods people wanted, leading to shortages within an inefficient centralized system. In transforming socialist societies switches from capitalism to socialism or *vice versa* are ideal occasions for cutting women down to size and moving the goal posts. The condemnation of 'capitalism' in the first place might be seen as just an excuse to abolish women's businesses, which were more numerous (although much smaller) than men's capitalist enterprises in Old Russia, because of ancient fears of women gaining power and transforming society on their own terms, possibly through witchcraft (Clements *et al.*, 1991).

Because the emphasis in socialism was on economic issues – and 'the higher cause' – social and cultural considerations which affected women's lives were marginalized. Public culture such as sport and performing arts was strongly supported, but there was little space for consideration of women's domestic concerns in a society riven by bourgeois/revolutionary dichotomies. Under scientific materialism everything, including women's position, was meant to be 'determined' by 'objective' factors. The former Soviet Union expressed a changing sequence of views about what women 'should' be like (Attwood, 1990; Buckley, 1989). These have oscillated between being pro- and anti-abortion and for and against traditional family structures. Women have been seen as either the same and equal, or special and different. In the former Soviet Union the production of children was seen as a public-sphere, industrial issue, and mechanized by the state, albeit not for the benefit of women themselves. This may partly explain the different agenda, mutual incomprehension and potential dichotomy between western and eastern European feminism (Lipovskaya, 1993). Much socialist guidance to women reflected attitudes derived from pre-revolutionary conservatism, folklore and tradition. Women were actively involved at the beginning of the Revolution, but were systematically excluded as time went on. History was rewritten and women were either left out or represented as drawing room socialists (Maxwell, 1990). But there were always a few prominent women for public relations purposes, such as Valentina Tereshkova, the first woman astronaut, who has apparently supported women's rights (Mamonova, 1989: 168). Women as a whole have been the first to be expected to make sacrifices when it comes to the allocation of scarce resources. This attitude is also a feature of western societies under war-time economies. Rationing may be imposed on the civilian (mainly female) population, but the men at arms are issued with meat whilst the women are starving (Adams, 1990).

Much has been written about the practical problems for women in socialist countries – such as having to share kitchens with several families; the lack of privacy in small apartments; the double shift without the benefit of modern domestic consumer goods such as washing machines; the lack of adequate shops and queuing for limited supplies of food and consumer goods; and the lack of birth control provision, except by abortion (especially China with its one child policy) (Attwood, 1990: chapter 8). Benefits such as childcare provision were always somewhat patchy, but are now being withdrawn to force women back into the home, and give 'back' their jobs to men, although women are frequently the main breadwinner (Heinen, 1992). An economy centrally planned by men meant that there was a surfeit of tractors and no sanitary towels (Drakulić, 1987). In Britain sanitary items are subject to VAT, which suggests they are seen as a private luxury rather than a public necessity. Whereas in the Soviet Union anti-consumerist attitudes resulted in long queues for essential goods, in Britain it resulted in town planners ignoring the needs of women for shops on housing estates and for local opportunities to develop flexible forms of employment. There were some advantages for women in socialist societies, such as a puritanical public attitude towards women and no pin-ups or pornography – something which changed rapidly with glasnost. But there has been a long tradition in the Russian language of verbal pornography (called мать *mat* (mother), Mamonova, 1989: chapter 16). This can make women feel as out of place in public areas as more visual manifestations can in western countries.

Population planning and eugenics

If economic and social planning seem inadequate means of control, one can go one step further and control the supply of human beings which are needed for the plan, so there are not 'too many people'. Population control is the area of planning in which the individual is most assumed to be irresponsible, and women often get far more blame than men, although it takes two and women are not necessarily the willing ones. Such attitudes are prevalent in certain branches of town planning. There is a certain mentality which is highly condemnatory of women for being selfish and having either too many or too few children. It can be either. The debate is likely to be conceived in abstract terms, with arguments based on public-realm mathematical predictions that ignore personal and cultural aspects. Because policy making is developed in a formal, impersonal setting within the public realm, the planners may not bother themselves with all the messy issues of contraception, abortion and rape which constrain women's behaviour in the so-called private realm of inter-personal relationships. Such women's issues might be dealt with by quite different departments from those dealing with strategic population planning issues. Mackenzie's study (1989) juxtaposes family planning and town planning issues to shows how women have to negotiate with a range of state agencies to get what they want.

Although policies are abstract and mathematical, and perceptions of women generalized and stereotyped, the planners can be very 'personal' in their attitudes towards individual women (Greed, 1991: 116). Many women planning students and practitioners have had to put up with volumes of sexist comments, innuendo and condemnatory statements over the years, interspersed in college lectures, office conversations and committee meetings. Seesawing: it would seem that women planners of the 'wrong' class or ethnic origin are picked on far more than middle-class women, who may not have experienced these problems to the same degree. This reflects the acceptable/ unacceptable women dichotomy (p.12). Men are possibly less likely to pick on a young woman who they know is the daughter of a chief planning officer, but they will readily do it to women who appears to have no boyfriend or influential male relatives in local government or the professions to defend them. Some women may not realize they fall into a protected category and therefore may be quite condemnatory of women who do experience harassment, as if they brought it on themselves. Such 'protected women' are the ideal ones for promotion under equal opportunities policy, which usually operates with little overt acknowledgement of the part class plays in the selection process. Boundaries related to class as well as gender are emitting force fields in such situations. Many of the dichotomies which are associated with male/ female differences such as pure/impure, clean/dirty, mind/body must be applied to perceptions of middle class/working class women to make sense of why some women are seen (and see themselves) as more acceptable than others, as women, too, have been divided along class lines. You must live on the right side of the railroad tracks (and the dichotomous divide). The disadvantages of being the wrong gender may be compensated for if one is the right class (Greed, 1991: 132).

Population planning links across to the worlds of medicine and social policy, where arguably the most powerful and patriarchal planners reside in our society today (Oakley, 1980; Savage, 1986; Greed, 1993b). Modern family planning might be seen as continuation of nineteenth-century eugenics, which is the science of the improvement of the human race through selective breeding and racial hygiene (Mazumdar, 1991). Spallone (1989: 143) points out that the *Eugenics Quarterly* became the *Journal of Social Biology* and the *Annals of Eugenics* became in 1954 the *Annals of Human Genetics*. Reproductive technologies are presented today as a way to help the infertile (Spallone and Steinberg, 1987). The success rate is very low. The real reason may be to pursue genetic engineering to create the ideal human being, and eventually to replace women with an alternative maternal environment (Stanworth, 1987). There is a class bias in family planning; working-class and ethnic-minority women are more likely to be sterilized, and to be told, 'people like you always have too many children', third-world women have questionable contraceptive drugs dumped on them from the west. Some white, middle-class women are given every help to solve infertility problems, but they are not necessarily any more powerful in controlling their own reproductive rights and may feel pressurized to have children (Spallone and Steinberg, 1987). But professional women with minds of their own might also be seen as uppity and therefore unfit. Unfortunately, many of the personnel involved in making these biased judgments as to who is fit to have children or not are other women working as medical personnel or as scientific staff in reproductive technology establishments. More women in a specialism does not necessarily mean 'better' for other women (Greed, 1988). This is echoed by Mies (1987: 41), who states that demanding 'more women' in the area of reproductive technology is short-sighted, because we must ask what policies and aims these women hold. Many patriarchal medical women are completely socialized into the male professional culture. Such women, although planners, are not zone zappers but assistants to male zoners, and key contributors to the maintenance of patriarchy; presumably in return for some personal sense of superiority and power over other women. Mies states that technology is an instrument of domination even if women control it (cf. Firestone, 1979). First-wave urban feminism was strongly linked to the eugenics movement, another controversial issue because of eugenics' subsequent associations with fascist governments, death camps and selective breeding programmes (McLaren, 1978: chapter 8).

Third-world planning

Town planning has long been linked to population planning, eugenics and colonialism, these movements structured around pure/impure, order/chaos dichotomies, as examples from *The Planner* over the century illustrate. In January 1953 (Vol. 39, No.2: 34) an article by R. A. Jensen, 'A national planning enigma; high density, overspill, or emigration', endorsed the emigration of working-class people to Australia, and prefigured some of the more sinister debates within the present-day 'green movement' about population polluting the planet. A review (Vol. 42, No.2: 46.7, January 1956) of *World Population and World Food Supplies* by John E. Russell (1954) points towards the future global aspirations of town planning. 'Scientific' town planning of the 1960s strongly espoused population control measures, and nowadays echoes of this are resident in green and third world planning. We have hardly begun to come to terms with the links between planning and colonialism (Ware, 1992: 169) and how this has affected the way in which ethnic minority women are perceived in Britain. Rakodi (1991) comments that the women and planning problems are not particularly different in the Third World from those in the west but rather they are more intense.

The same gendered dichotomies inform the way problems are perceived and how planning policy is framed. Problems derive from attempts to superimpose western gender roles and public/private dichotomies on other cultures, so that the men are seen as farmers and women as housewives. The literature is full of examples of planners applying high-technology solutions to agricultural problems, whilst ignoring the fact that women are the main farmers and will now have to walk twice as far for water because that new wonder-dam has taken away their existing water supply (Rogers, 1980; Momsen and Townsend, 1987; Moser and Peake, 1987). Also, the western experts are far more likely to ask the village elders than the women themselves what is wanted. Creating new public spaces through western-type central area planning may appear to benefit everyone, but such areas may be commandeered by men if, according to local cultural attitudes, women are not be allowed to move freely on their own in public space. Women may be jostled on public transport if they are seen as taking up men's space in countries where there are few private cars. They may feel safer wearing a veil, taking their private covering into the public realm (Benzerfa-Guerroudj, 1992; Lateef, 1990). From some middle-eastern countries one hears tales of separate open space, footpaths and buses for women, but also of women being forbidden to drive cars. Women may get the worst of two worlds when their own traditional gender roles are threatened by progress, and the roles given to them by their western 'conquerors' are far worse. The establishment of military bases, and modernization schemes to attract the tourists, whilst possibly improving the local economy, might destroy traditional ways of life and destine local women to become maids and cleaners, or prostitutes (Enloe, 1989).

Other dichotomies which particularly disadvantage women in the Third World are those between clean/dirty and between order/disorder. Because of the emphasis on health programmes in third-world aid initiatives, town planning has often reflected a concern with cleaning up the city: rooting out disease by bulldozing the shanty town; getting rid of the clutter, and 'modernizing' the situation; trying to get the city to conform to some sort of sterile western ideal. Zoning policies introduced in the name of hygiene may make it very difficult for women to combine home and work activities if they are not allowed to graze their animals, plant vegetables, or carry out crafts around their homes (Rakodi and Mutizwa-Mangiza, 1989). Agriculture in the form of small-scale 'husbandry' appears to fall foul of the public/private dichomoty falling into the female/domestic/dirty sphere. Large-scale male-controlled agriculture is usually seen as a serious, clean, public-sphere activity. The introduction of modern sanitation and basic sites and services schemes can create further problems for women, if a gender-blind preoccupation with 'germs' predominates over any sensitivity to local customs. For example, it may be considered immodest by women to use public toilets which are built with a gap between the floor and the outside wall (for ventilation presumably). Quite simple things like putting water taps too low for women to fill up their tall water jars need to be taken into account (Rakodi, 1991). Likewise, when designing for the needs of ethnic minority groups in Britain different ways of doing household activities have to be borne in mind. The 'Jagonari' Asian women's centre (at 183 Whitechapel Road, London E1) built by MATRIX has ground-level sinks for washing large cooking utensils.

Village women who have developed their own self-help housing groups and challenged official policy could be seen as zone zappers (cf. Moser and Peake, 1987: chapter 7 on Nicaraguan women). But, third-world planning is often associated with large salaries for the planners, who are solving 'world poverty' by working for the World Bank and other similar organizations. In contrast there are also plenty of volunteers who do sacrificial work for nothing or for low salaries, such as aid workers and traditional missionaries. What often happens is that the people themselves are the first to campaign about a problem but no one hears them. Then, overnight, they find that their problem has been taken over and turned into a source of salaries, careers and academic reputations by the predatory professional classes. This is echoed in the west in the race-relations business, the incest industry and homelessness campaigns. The problems which women suffer might actually be used against them, to increase patriarchal power. Having distanced themselves from the people, and lost potential goodwill, the third-world experts might lauch a public participation exercise. If this receives a low response it might be put down to apathy on the part of the local people rather than arrogance by the western workers. One of the problems is that there are too few women planners working in the Third World who might act as a link with third-world women. This is especially so in societies where the sexes are segregated. Western women used to be excluded on the pretence it was too hot for them, although more than half of the indigenous population are women, and white tourists go to hot countries for the sun. The problems of women not being involved in the decision making process is replicated countless times in Britain in the location of shopping centres, community facilities and the priorities given to male journeys in public transport.

Green planning

Having sought to dominate nature, now man is now turning his mind to saving it, because doom will result if nothing is done! Sometimes one would hardly imagine that people actually have the right to live in some of the areas which are being saved, in particular local women, who are likely to be blamed for cutting firewood from the rain forest. Unfortunately native peoples might not be seen by man as being as important as other manifestations of nature like endangered animals or the environment. In parts of Africa they are banned from farming or hunting in their own tribal lands, which are now designated as nature reserves. Women often appear to fall down the gap in the middle of the dichotomy between man and nature. Many of the problems women encounter *vis à vis* the environmental debate derive from the fact that they do not quite fit into either category. But it is generally accepted that women such as Rachel Carson with her book *Silent Spring* (1965) started the green movement. Women such as Nan Fairbrother (*The Planner*, Vol. 57: 398, 1971) and Marion Shoard (Vol. 68: 4, 1982; Shoard, 1980, 1987; Griffin, 1984; Caldecott and Leland, 1983) have been at the forefront of promoting green issues in and outside town planning, acting as zone zappers to champion nature against malestream science. Several of the women who espoused green values subsequently showed themselves to be feminists too, for example Janet Brand, who in 1972 had written in the journal about 'the rape of the earth' (Vol. 58: 172). Hilary Coleman (Vol. 66: 45, 1980 – not to be confused with Alice Coleman, 1985) did much to raise the profile of the importance of the bicycle in planning policy. Jennifer Armstrong is famous for holding the fort all the way through the Sizewell Nuclear Plant Inquiry in 1983. Nowadays groups such as WEN, the women's environmental network, continue to promote a linkage of green and feminist issues, but still the malestream does not appear to see the connection between environmental sustainability and treating women as human beings. The Labour Party's report on the physical and social environment (summarized in *The Planner*, September 1988: 11) barely mentioned women at all, just children, and Chris Shepley commented (p.15, ibid.) that this document was 'less sexy' (sic.) than expected. Somewhat 'hippy' 1960s groups such as Friends of the Earth have changed beyond all recognition over the last twenty-five years and now have respectable, Establishment-type male leaderships, as the organizations move from the private to the public arena. Some commentators prophecy the complete takeover of nature by man: a 'Biotech Global Takeover' (Shaw, 1991). Through science, humans will be reduced to mere 'cyborgs' with the scientist acting as God, planner, super-technologist and evolutionary agent (Haraway, 1991). It seems to me that the environmental movement could even provide a route for man to achieving imperialistic, patriarchal power over the planet. It is one example of a trend I have observed in which, far from the public and private zones existing with some respect for each other as separate spheres, the representatives of the public, male zone seek to take over, control or even destroy the private, female zone.

Both 'women' and 'urban feminism' are being marginalized by aspects of the green movement, in particular being upstaged by sustainability (Blowers, 1993). The times I have been told, 'we have done women, you should be concerned about green issues'. I do not like 'greenie' public planning policies which only work because women have to do more private domestic work. Hinnells (1991) suggests that shoppers (women in 80 per cent of cases) should carry back their waste packaging for recycling. House to house collection would be preferable. Likewise the Richmond draft plan (LWPG, 1991: 14, Policy 4.142) recommends microcycling, which is commendable provided women don't become unofficial 'dustmen' for everyone else's rubbish. Some manifestations of the ecology movement, and animals rights activism, are positively misogynous in blaming women for having too many children, being consumers, using washing machines and wearing cosmetics or the wrong clothes. This attitude disregards the fact that they are not free agents, and much of what they do is for others. As Kitzinger (1991) points out, the anti-fur coat advertisement slogan, 'it takes up to 40 dumb animals to make a fur coat but only one to wear it', legitimated hatred of (rich) women and did little to help the campaign. In fact many women are in the forefront of campaigns against violence towards animals, including farm animals: most of whom are, themselves, female (cf. Collard and Contrucci, 1988). Because the environment is a relatively new area of debate, some women planners saw their opportunity and succeeded in becoming consultants (Fortlage, 1990). It will be interesting to see whether they retain their lead. See-sawing: recently, I received a consultation document on EQ (environmental quality) which looked at first as if it were on EO (equal opportunities). But the acronym had purposely been 'masculated', by the addition of a 'tail', deftly shifting emphasis in the built environment professions from 'women' to 'green' issues, and from feminism to sustainability.

EQ is not to be confused with QA (quality assurance), which is an unwelcome governmental imposition to maintain standards in the construction profession. QA, as one woman in the 'Women in Design and Construction' group pointed out to me, (unlike EQ) is becoming a classic woman's area of expertise marginalized from the mainstream of professional work (the 'give it to her, nobody else wants it but we've got to do it' syndrome (cf. Greed, 1991: 138–9). Since planners talk in acronymns it is vital to crack the code.

DESIGNING URBAN REALITY

Designing for whom?

In the past, the planner featured primarily as artist or designer, rather than as urban manager, scientist, or social reformer, but, as will be seen, the beliefs he brought to planning reflected attitudes to women's place and role similar to those of his more scientific modern brethren. To the artist planner the city has been seen as symbol of civilization and place of culture (Fisher and Owen, 1991), and not just as a scientific system, a property portfolio, or a collection of social and economic factors in need of control. The ideal of the planner as artist is still to be found today, as expressed in the definition of the objectives of the Royal Town Planning Institute in the *Directory of Official Architecture and Town Planning* (Brett, 1989: 313):

to advance the study of town planning, civic design and kindred subjects, to promote the artistic and scientific development of towns and cities for the benefit of the public, and to promote the general interests of those engaged or interested in town planning.

Urban design is making a comeback partly because of the influence of urban conservation groups, Prince Charles (1989), and the town planning lobby with the RIBA. Planning can be seen as a non-political, non-ideological activity, purely concerned with civic design, as if one could separate the spatial end product from the social and economic processes which created the demand for the architecture and townscape. This attitude is often linked to a disdain for ordinary people living in the designers' wonderful creations, and of their practical needs. This reflects the influence of public/private dichotomies in the way the planner sees the city as public space. The *beaux-arts* approach has tended to see women as merely 'figures in the landscape' (Dresser, 1978, referring to Davidoff), and dwellings just as stage sets with no backs to provide somewhere to put the washing and for the children to play (Roberts, 1991). Likewise much post-modernist architecture is based on façadism, and creating a false reality.

One senses a desire by the artist planner to please some invisible all-seeing Divine Being in whose eyes beauty is judged with little consideration of convenience for the residents. Such attitudes are still latent in modern town planning. See-sawing to the present day: in the case described in the article, 'Hot food takeaways and planning control' (*The Planner*, 2 August 1991: 7–8) the amount of litter likely to be generated by customers appears to outweigh considerations of why more 'people' want take-away food nowadays; perhaps because their wives are working? Not only are such attitudes classist, they have also been found to be racist when refusal rates for 'ethnic' takeaways have been compared with those of white, sit-down restaurants. This is a matter of concern to the planners (RTPI, 1993; and 'A blind eye in the town hall?', *The Guardian*, 26 March 1993: 19 of supplement). In the same vein maintaining urban 'beauty' might involve such sexist policies as getting rid of unsightly bus shelters and not replacing them, or putting shrubbery around a women's public lavatory to create a mugger's paradise (or closing it altogether because of vandalism). The 'designer' designs by looking 'at' an area, not by being 'in' it as a resident who 'uzes' the area in all weathers.

Designing by whom?

I would now like to consider the gendered nature of the preferred qualities of the artist planner, the genius who can have such a powerful role in creating urban reality. One of the requirements for being recognized as a great artist planner is the ability to produce good design, but this is not quantifiable, like traffic volumes and office space, since it is related to immeasurable qualities such as good taste, aesthetic sensibility and intuition: highly gendered attributes. These are the very qualities that women have been condemned for possessing, because they are signs of subjectivity, levity and weakness. But, when men possess them it is a different story. As Battersby (1989) points out, only a male can be a 'genius' in our civilization. To be a genius he must possess 'female' qualities to an even greater degree than women, but will not, necessarily, be condemned for this. Such a man will usually be preferred to a woman, even a patriarchal, sensible woman. Artistic qualities are apparently innate and, like practicality, cannot be acquired by education. You simply have to be the right type and visible within the right context to start with, reducing the pool of potential people from which to choose. A person's 'natural' gift must fit into the expected standards and fashions of the time. In the past this often entailed having an intimate knowledge of the orders of classical architecture, and one could not learn these unless one was articled or apprenticed to a Master, and to qualify for this one had to be male. As a result of participating in a working party dealing with assessing competence (for NVQ – National Vocational Qualifications – purposes) in the construction professions, with particular reference to architecture, civil engineering and town planning (CISC, 1992a) I was to discover that some architects believed that 'creativity' *could* be measured (CISC, 1992b), the criteria being the approval of one's peers (mainly male) and the standard of the outcome of one's creativity (dependent on getting commissions). The quality of *intuis* (wisdom) was highly valued as the mark of a true professional in CISC discussions, above *cognis* (knowledge) and *technis* (practical ability), although, women, as against architects and civil engineers, are usually condemned for their intuitive qualities (Macleod, 1991). Etymologically, the engineer is the one inspired by genius; compare 'Femmes en genie', the French women-in-engineering group.

A person may fail to reach the professional world because of gendered expectations going back to junior school which affect the chances of a person being discovered or recognized. A creative, white middle-class boy might be encouraged to be an architect but a creative girl might be directed towards art college. If a black child is creative in art, this might simply be explained by the fact that 'they' like bright colours, or if creativity manifests itself in music it might be explained away by the fact that 'they' have natural rhythm. One wonders if Stephen Wiltshire, the young, male, black artist who has achieved international fame and admiration from the art world for his drawings of buildings and urban skylines (Wiltshire, 1989), had not also been autistic whether he would have been spotted. Paradoxically, possessing more than one minus minority-factor can put a person in a more advantageous position, a point made by black women barristers (*New Law Journal*, Vol. 142, No. 6554: 746–8, 29 May 1992). The concept of the Artist as a gifted individual and genius who is the possessor of a special vision derives from the Romantic movement and was built upon in Victorian times (Abrams, 1953; Babbit, 1947; Peckham, 1970). Whilst some men were being given star status, established women artists were increasingly marginalized, their work demoted to the secondary realms of craft rather than art (Anscombe, 1984). The man who went on the Grand Tour of Europe might return as an architect; the equivalent woman would return 'only' as a cultivated traveller, and possibly a fallen woman too. Several authors have catalogued the marginalization of women in art (Greer, 1979; Spender, 1982; Chadwick, 1990; Broude and Garrard, 1982). The idea of individual male creativity was bolstered in the twentieth century in popular books and films, such as in Irvine Stone's accounts of the lives of Van Gogh and Michelangelo who were portrayed as mad, macho figures (Stone, 1961). The image of the fiery, bearded, eccentric male genius affected twentieth century concepts of the real architect, town planner and socialist. Nowadays it is not enough just to be good: to be recognized as great as an artist or architect one must have a suitably packaged biography (Wayte, 1986 and 1989) so that a couple of 'squiggles' on a piece of paper can be interpreted correctly.

One can see a star system at work in the built environment professions – especially in architecture, surveying and town planning, which generally only create famous men. Young women are often given the impression they cannot be architects or planners because they lack the right technical or mathematical skills. In reality many architectural schools will not accept technical drawing as a valid academic subject, and although a basic mathematical ability is required, colleges will welcome students with arts A levels as a sign of general cultural knowledge. Some will welcome Art A level and evidence of creativity (and not judge it as a non-subject) provided it is balanced by evidence of overall academic ability. One senses a double standard because young men of the right type, who will eventually grow into alpha-males, might be allowed to show more artistic (even intuitive and spiritual) and less mathematical ability than young women, who may be expected to fulfil all the official criteria. Few women seem to succeed as artists in the world of architecture and town planning, but some men do. For example, Francis Tibbalds, who, sadly, died in 1992, was one of the youngest Presidents of the RTPI, and his penmanship may be seen illustrating Heap's work (Heap, 1991, who was also coincidentally a past president of the RTPI). Paradoxically, the mathematical woman is one of the types of woman most likely to succeed in modern town planning (see chapter 10). Some might argue that women should not be concerned about achieving male-defined forms of success, but in professions where the star system operates strongly it would seem that an individual has to achieve fame and recognition in order for his/her ideas to be taken seriously and thus have any influence.

In the twentieth century scientific quantification *inter alia* takes over from spatial design as a driving force in town planning (where, as will be shown, density is the spinode, p.116). But the detachment of the grand-manner scale lives on reincarnated in certain modern branches of planning such as British state house-building programmes, French *grandes ensembles*, North American transportation planning, and a general enthusiasm for large-scale abstraction expressed in such politically diverse realms as urban systems theory (McLoughlin, 1969) and neo-Marxist urban conflict theory (Pickvance, 1977). A God's-eye view to urban design persists in some quarters – the town planner looking down on everything from above as 'seen' set out on a drawing board or in a model. Architectural applications of 'virtual reality' software, CAD (computer aided design) and GIS (Geographical Information Systems) reinforce this illusion and create a safe, controllable, artificial reality which renders invisible what the viewer does not wish to see all around 'him'. No wonder some modern, technologically advanced men planners can (without guilt, whilst being very pleasant as people) completely ignore women's demands and claim sincerely that they support equal opportunities. They have not 'seen' women, as their view is within the confines of another, separate realm of (false) reality which is complete and all-encompassing to their eyes. Significantly, the size of the women's toilets provided in CAD packages for office layouts are far too small to be practical. Size matters.

5

REFLECTIONS ON THE
HISTORY OF PLANNING

THE STORY OF PLANNING

Expectations and Realities

In this chapter I will reflect upon how historical town planning has been
shaped by beliefs about how the world ought to be. Such beliefs, especially in
the need to separate the sacred from the profane (Eliade, 1959), have been
reflected in the layout of historical cities, sometimes in the creation of
religious buildings and spaces, but also in spatial divisions between male/
female, public/private, elite/masses for a variety of more prosaic reasons. The
creation of urban form is all about imagination (and suspending disbelief). The
planners act as 'imagineers', as the designers of Euro Disney are called, (Sky
TV, 12 April 1992, News feature), the creation of reality being a function of
religions (Eliade, 1959). In passing, although some see Euro Disney as a
cultural Chernobyl, it is one of the few new towns in Europe designed with
the imagined needs of families with children in mind, but where, I'm told,
people are not allowed to take in their own sandwiches. First, I will reflect
upon how the subject of the history of planning has been used to create a
patriarchal image of the past to legitimate women's place in the city today.
Then I will consider the different historical periods. But, my purpose is not to
give an (in)complete historical account, as one can easily be drawn into a broad-
brush discussion of the creation of civilization which, by definition, is
essentially urban (cf. Clark, 1969). Rather, I will draw out selected examples
which illustrate my themes of belief, zoning according to perceived dichotomies, and the role of women as zone zappers transcending and transforming
such divisions. However the sequence is broadly chronological.

In studying the history of planning, both as a student and lecturer, I
experienced the same disappointing credibility gap between expectations and
the realities of the nature of the subject matter. Again, the problem is that the
subject has generally been developed from the point of view of those who
reside in the public realm, often within the town planning profession itself:
that is, within the world of men (literally for many centuries). But, I cannot

help my own personal, domestic, feminist and non-professional perspectives creeping into my judgment of urban phenomena. I juxtapose material from these other realms with the mainstream account in order to disturb and unsettle the discourse. The history of planning is a component of town planning education, and might be seen as part of the rite of initiation into the town planning. I have chosen to focus upon the history of planning, but one could undertake a similar analysis of the social role of any of the other main subjects on planning courses in the process of professional socialization. I am aware from feedback from female colleagues that some women react quite differently to aspects which I found distressing. This raises the question 'is it just me?' (a common worry among feminist researchers). I am simply giving my reflections as informed by life experience, and feminist scholarship, for the reader to think upon.

Readers who are unfamiliar with the history of town planning should refer to the mainstream texts listed in Appendix II, pp.197–8. Some of the books appear quite innocuous and useful, until one asks 'where are the women?' and 'what has been left out?' and 'what assumptions are they based on?'. I will discuss the earlier periods in more detail because these frequently set the tone of the subject in the mind of the student. Guidance on feminist critiques of the history of planning are given in Appendix II. There is also the middle ground of enlightening urban history books by women which the authors may not see as feminist in intent (such as Darley, 1978, 1990; Rosenau, 1983; and by Alison Ravetz, 1980, 1986), but which men may see otherwise. A male planning lecturer advised a student, 'Ravetz? that's really a book about women and planning, I'd go for something more solid if I were you'.

'Other' dimensions

Not only have feminist interpretations been excluded from the official account, but other hidden themes and subterranean streams have been edited out, particularly those of a spiritual nature. Many planners and architects in the past were variously spiritual, necromantic, kabalistic, and geomantic in their approach to designing individual buildings and city plans. They sought to incorporate cosmic harmonies (cf. Stewart and Golubitsky, 1991; J. Greed, 1978 on ley lines) or to please the spirit world and the dead. Religion was part of the public realm of men, giving them divine rights of superiority, for much of history. Over the centuries, religion shifted sides across the dichotomy. By the twentieth century, aspects of religion and spirituality, especially personal devotional activity, became more commonly associated with the private realm, and especially with 'superstitious old women'. But, modern civilization is still based upon strongly patriarchal, religious beliefs about how the world should be, albeit 'ungodded', and manifested in secular humanism and scientific spirit.

The overtly religious aspect of town planning was swept under the carpet in the twentieth century, when writers offered alternative economic, biological or anthropological explanations of urban design from a broadly humanistic and scientific perspective. This trend reflected the secular/spiritual dichotomization within society itself. Perhaps today some feminists are more successful and acceptable than others in the world of town planning because, beneath their feminism, they hold the beliefs which coincide with those of modern scientific planners, whilst others hold to unacceptable superstitions which do not fit. It would seem that women planners with certain 'spiritual' tendencies and beliefs are unlikely to be considered the right type for local authority employment. These beliefs are far more of a minus factor than those of a material, socialist, feminist. One of the reasons men have sought to use civilization to control women is, arguably, because of fear of women's spiritual powers. Woman is truly the unknown quantity, the sphinx in the city (Wilson, 1991). Currently there appears to be a paradigmatic shift occurring towards matters of the spirit, as manifested in the New Age movement. But supernaturally inspired planning is not necessarily more positive for women, in spite of the enthusiasm for goddess cults in some feminist circles. Although the female element might be recognized as powerful, it might still be seen as potentially dangerous and in need of control by man (cf. Walters, 1989 on Chinese *feng shui* planning, which reflects yang/yin (male/female) dichotomies).

The past justifying the present

One function of history, in general, is to justify the nature of the present as being normal. It is hardly suprising that it is 'his'story, rather than 'her'story which has been the focus of attention, as history is normally written by the victors. In passing, ironically, the word *historia* (from 'ιστορέω, to inquire) has a reflective, ethnographic, almost feminist set of meanings including knowing, judging, story-telling, inquiring and searching. Indeed, studying history may be seen as 'listening to the dead'. But, 'malestream' history is not a reliable subject; like statistics man can prove whatever 'he' wants through historical research. Like sociology it is value-laden and open to intepretation. Philip Abrams, the urban historian, has commented that 'history and sociology are, and always have been, the same thing' (Abrams, 1992). Not only is the mainstream account gendered, but women historians offer different feminist accounts of what really happened, reflecting their own sociological perspective on history.

The teaching of the history of town planning is one means of transmitting the mores of the planning tribe to the next generation, socializing would-be planners into the values of the professional subculture. In comparison, many surveying courses put far less emphasis on history, and students certainly do not need a knowledge of their history to qualify as surveyors. This is because their subject is seen as practical and commercial in nature (Greed, 1991: 35), whereas planning is more academic in content and image. This impression is reinforced by the fact that prominent planning historians, such as Gordon Cherry, have also served as Presidents of the RTPI, and that many of the introductory textbooks on town planning are really history books (Ravetz, 1980, 1986; Hall, 1989). But, town planning – compared with architecture, surveying or civil engineering – has a rather shaky pedigree, and limited special technical content. To justify their existence as a distinct profession, town planners make much of their illustrious past. This is similar to the borrowed glory factor identified in my study of surveyors, who were more concerned with establishing their present-day class credentials by associating themselves with other higher professions, such as law and accountancy (Greed, 1991: 59). Ironically, many of the earlier manifestations of planning had nothing in common with the state socialist planning which emerged in the modern period, but that inconsistency, is conveniently ignored.

The way in which the subject and meaning of history has been conceived has been highly gendered. Women have questioned many aspects of men's interpretation of history, in which progress for mankind is often limited to progress for white European males. Students are told 'how bad it all was', especially for the (male) working class at the time of the industrial revolution, before society's saviours, the modern planners, arrived on the scene. But for women the industrial revolution did not bring such great change, or progress. Indeed, women's domestic work underwent only superficial change in the last century (Zmroczek, 1992). Whilst man has been besotted by ideas of ascent and progress (Malinowski, 1977) women historians have dwelt on the descent of woman (Morgan, 1974). Carr (1965: 111) warns about seeing history as a science concerned with proving the progress of human affairs. Popper (1958) draws attention to the subjective nature of history, and the danger of adopting a determinist approach based on speculative sociology. Bebbington in his fascinating book, *Patterns in History*, takes a cautious view of historicism from a spiritual, proto-fundamentalist perspective (1979: chapter 5). Although planners may subscribe to a determinist view of history, they have no qualms about helping its process along, otherwise they would be redundant as state interventionists.

PERCEPTIONS OF THE PAST

Civilization

Studies of early civilizations make much of the forces of geographical determinism in the creation of the first cities in the Middle East, stressing the importance of the need for a surplus of food so that people (men) could apply their energies to uplifting activities and create civilization. Many criticize the gendered (and classed) conceptualization of 'work' therein, for 'a woman's work is never done'. The creation of a parasitic, luxury class (Veblen, 1971) usually resulted in masses more work for the slaves; Homer states that most slaves were women (Eskapa, 1987: 23). Their services went beyond modern definitions of work: slavery and sexual slavery being linked from ancient times. Feminist scholars question the underlying definition of progress in this view of early history. Why, when early man no longer needed to work so hard, didn't he simply sit down and have a rest? Why did he possess this striving to build cities and civilizations? (as Morgan asks, 1978). Civilization, like capitalism, can never stay static. It always has to grow and requires a constant supply of new resources, manpower and land. In contrast, some so-called primitive and under-developed civilizations (Knappert, 1990; Bushnell, 1968), such as the aborigines and unlike western man, built no cities, but meticulously named every place and apparently lived in harmony with the Earth. How, whether, and when the shift from matriarchal to patriarchal power came about is open to much debate but feminist explanations are no less valid than Mumford's version (1965) (as he was not alive then either). Archaeological research has been undertaken to substantiate alternative feminist interpretations, for example by Lucy Goodison (1990). On the basis of her research findings, she seeks to completely reconceptualize the whole history of the world. She questions the 'male' defined dichotomies between the material and the spiritual, and between the intellectual and, the natural, and by association, between the male and female as human and non-human upon which the concept of civilization rested. The starting point of 'civil-ization' is perceived negatively by many feminists as the time when priests, kings and philosophers convincingly established formalized male authority over society, displacing women's more spiritual and natural power (Miles, 1988), and thus consigning women's attributes to the zone of the 'other' on the wrong side of the dichotomy.

But, women, like men, can fall in love with deterministic views of history. Some feminisms, especially liberal feminist versions, imply that everything will get better once society acknowledges the problems (Tripp, 1974). Others take a more negative view, arguing that feminism is on the way down (Segal, 1987) and that a backlash has set in (Faludi, 1992). One's viewpoint on the possible course of history has important implications in judging the efficacy of the efforts and strategies of the present women and planning movement. It raises deeper questions about the limits of self-determination, freewill, and fate to which women in their individual lives so often feel bound. Some feminists give the impression that they see the last 5,000 years of civilization as a mere hiatus and look forward to the re-establishment of a new matriarchal age. This has parallels with Christian dispensationalist theory, and the idea of living in the tribulation period. There is nothing more effective than looking back to some ancient civilization and using it as the base to justify one's ideas of what society ought to be like in the future, a ploy used by a variety of historians, biologists, evolutionists, and an assortment of political theorists. Reference to a golden amazonian age, or some ancient goddess cult can be used to legitimate the claims of modern feminism, by arguing that feminists are simply re-establishing what was normal before the interruption of a temporary patriarchal age (Heine, 1988). Significantly, many of the female-future, alternate plans for how society should be contain an openly spiritual dimension which is not to be found in modern town planning (but was strong in utopian planning in the past, Kumar, 1991). It is notable that a mandala symbol appears on the cover of the RTPI women and planning working party report (RTPI, 1989b).

Personal perspectives

When I was a new planning student, at the beginning of year one we 'began at the beginning' on the history of planning. Speaking as a lecturer, we are lucky today if we have the time to begin from the Renaissance, what with the pressures of course modularization, but the same 'problem' is latent. Straight-away, I obtained the main textbook, namely Lewis Mumford *The City in History* (1965). I began at chapter 1 and was stunned by the content. I could not understand why it was so 'sexual', and prurient at that, with endless references to women's reproductive organs throughout chapter 1, when it was meant to be about town planning. It seemed so strange and inaccurate. Primaeval women were presented as ageless archetypes, which apparently explained the nature of the modern housewife. I had only previously found these sentiments expressed in the *Flintstones* cartoon. Other women have told me Mumford's book did not hit them in this way, possibly because they were either more knowledgeable or less sensitized to the issues. I was unsettled by his observations, never before having read books written in the style of 'men writing to other men' about those objects called women.

In an earlier book, *City Development*, Mumford (1945: 53–6) reduces women to mere sex objects, declaring that the eye of man can equally appreciate 'the flanks and fetlocks of a horse, or the flanks and belly and buttocks of a woman'. He suggests that women regret the lack of violence shown to them by modern man, describing the 'resentment on the part of the female for his [the modern male's] lack of really persuasive aggression': almost suggesting that women want to be raped. Rape is an expression of power (Brownmiller, 1975), in this case overcoming the entire urban population through town planning policy, the ultimate control of the other. Graham King (1991) quotes Mumford as saying 'we must restore to the city the maternal . . . for the city is an organ of love'. He interprets this benignly (and is aware of women's issues, King, 1992) but, set within the context of Mumford's other writings, this comes across to me as the statement of a dominant, potentially violent male. Mumford probably wrote more about sex in his books on urban civilization than he did about town planning or cities, which is perhaps why he is seen as such a wonderful alpha-male, philosopher king by some in the town planning profession. But, several women academic acquaintances have quite a different view of Mumford, telling me that, unlike others, he did at least mention the female dimension of urban history. In his day he was seen by other academic men as a radical outsider, even a bit of a feminist, and he was not respected as a genius until after his death in 1990.

The image of 'man the hunter' who brought in the food was given great importance, references to this being dotted throughout a range of books on early man including Mumford. Paradoxically, in other lectures one might be informed that 'woman the shopper' was just a lazy wasteful housewife, who spent all her husband's money. Mumford, elsewhere (1930s), described women's shopping (marketing) as just 'fun', clearly not making this link between the public and private realms himself. I had always taken my forays around the supermarket for Mum very seriously and definitely identified more with the aggressive hunter than with the soppy shopper! I was beginning to experience that strange sense of living in (at least) two realities. At the time I concluded, as with much we had been taught at school, that town planning knowledge had nothing to do with real life but one had to learn it to pass exams. But, as a student under this constant barrage, I increasingly disassociated myself personally from the category 'women'. I have observed this attitude in some of my own female students too. See-sawing violently, from a class perspective, those 'stupid housewives' the lecturers were always going on about, must, obviously, have meant the very same 'lazy, stuck-up, middle-class women in the suburbs who never work' whom the girls at school had also despised, not women whom I knew from my own South London background where everyone worked. Landau (1991) explains how in 1964, for example, 80 per cent of school children left at the age of 15 to start work, everyone's mother seemed to work, and most schoolgirls worked in shops on Saturdays. (North seemed to American readers beware. Until relatively recently it was unusual in Britain even for middle-class school leavers to go to college.) It took me a while to realize the lecturers meant 'me' (and all women), not just lazy ones. I had put a class rather than gender explanation on their statements. Nowadays there are more students coming into town planning courses with variations of working-class backgrounds; albeit still a minority. But much of the material in planning literature, which they will read, is still written in the style of being about the working classes, as if it is unlikely to be read by them. I know some of my students are acutely conscious of the alienating effects of this.

I discovered that there were feminist accounts of early history and anthropological studies of primitive civilizations which were written by women *before* the men's versions (cf. Wirth, 1938; Sjoberg, 1965). As a student I gave academic men the benefit of the doubt, because feminism had hardly touched British scholarship (well, certainly not me). I assumed, innocently, that nobody had yet 'done' women. Nowadays, I look back in amazement that the old men of the tribe were able to give such a sanitized, male-centred version of the history of planning when many of them must have lived through the first wave of feminism in their youth. When I reconsider

Mumford now it seems to me he was having an argument with Margaret Mead (1949) rather than the ghost of Marx – that is with cultural feminism rather than with revolutionary socialism. A generation of men, including Freud (1856–1939) (Freud, 1935), made a career out of reasserting the importance of the male world and of sex in order to cover over the traces of the first wave of feminism which had established the importance of gender as a cultural rather than biological determinant of women's lot. Mead's anthropological studies demonstrated that gender was culturally constructed and not based deterministically on biological factors, illustrating this in her study of Samoa completed in 1925 (Mead, 1966).

It seemed illogical that other material had been left out. Coming from a background which included various other influences (variously, religious, occult, pentecostal, artistic and free-thinker) I was totally confused by the lack of emphasis on moral, religious and spiritual issues in planning education, especially the lack of reference to the Biblical account of early history, and the Jewish contribution to the development of civilization in the Middle East. It seemed standard practice for the early patriarch 'to know his wife, begat a son, and build a city' (Genesis, 4: 17). The Bible is a source of both positive and negative images of both women and cities, open to endless interpretation and cultural application to legitimize all types of mainstream and feminist viewpoints. Where reference was made in planning literature to Biblical sources the image chosen was frequently negative. The city is often presented as a sinful woman, for example Ashworth (1968: frontispiece) quotes from Zephaniah, 3: 1, 'woe to her that is filthy and polluted, the oppressing city'. Women, cities, and the urban masses, are often severally and jointly seen as filthy and unclean in certain branches of both town planning and theology. An anti-city theme may be observed running through aspects of the history of town planning, manifested in the desire tightly to control 'her' development and activities, or even to undertake 'the rape of the city' through violent demolition. Such ideas clashed with my ideal of the city as a shiny New Jerusalem, which had been influenced by images from *Pilgrim's Progress*, the story of a life-long journey to a celestial urban utopia (Bunyan, 1678). There are more positive beliefs which emphasize the purity, even spiritual nature of the city, as a form of paradise, as a place of refuge whose walls keep out evil spirits and the wrath of God, as well as earthly enemies (Eliade, 1959: 49), and as a place of joy, opportunity and creativity. Zephaniah is misquoted: the context implies his condemnation is against the male rulers of the city, and 'she' is used (conveniently) because 'city' is a female noun. It is unfortunate that Zephaniah's words were so mistranslated, for he may have been one of the few black prophets in the Bible (Newsom and Ringe, 1992: 225, refer to Zephaniah, 1: 1 as to his ancestry, and his links with Cush and Ethiopia). So few black people are mentioned in the history of planning.

Other others in Egypt

There seemed to be little inter-relationship between the planning version of history, and accounts from Bible study of the Jews' captivity and forced building activities. However, the phrase 'bricks without straw' (Exodus, 5: 11) stays with me as an apt description of 'EO policies without resources'. The Bible (especially the Old Testament) is full of references (see any concordance, such as Wigram, 1980) to spatial organization – that is town planning – including details of building regulations, temple design, land use zoning, rules for the location of boundaries, walls, fences, landmarks, places of refuge, plus descriptions of different types of cities and their functions and the nature of their suburbs, imprinting on space the laws of separation, holiness and order.

The 'reality' of the history of planning has been created with little reference to women and black people, with a concentration on DWEMS (dead white European males). Early civilization is normally seen as beginning in the Middle East and flowering in Egypt, where people were deemed 'civilized' and therefore white, and part of the Mediterranean cradle of civilization. In a reaction against the white male domination of history, Egypt, as prototypical civilization, has been claimed by women and black people (not mutually exclusive categories). But so-called 'minorities' can fall into a trap of claiming that they were really the inventors of some valued aspect of mainstream culture, or civilization which itself proves to be nothing more than a fabrication imbued with patriarchal, and often racist, values. Luomala (1982) suggests that Egypt was a matriarchy and the royal line passed through women. However the Egyptian form of urbanization was hardly woman-friendly, and even today Cairo is not an ideal city for women (Wikan, 1980). Society was strongly divided by class, with perhaps 5 per cent of the population comprising the ruling elite which included some powerful women, a small middle class, and vast numbers of slaves and agricultural workers. Class differences among women at various times in history can be potentially as significant as divisions between men and women. If women were powerful members of the Egyptian elite must they too take responsibility for the poor social conditions of the masses? Their priorities might have been quite alien to us. Urban planning was aimed at pleasing the gods and the dead rather than providing, for example, housing for the living. Compare the high cost to the living of *feng shui* burial layout planning in the Far East today (Walters, 1989). The veneration of the dead was prioritized over the needs of the living, suggesting a very different set of priorities from today (or does it? *vide* DWEBS).

Some argue that the ancient Egyptians were black (cf. Sertima, 1985) and that European civilization was an offshoot of 'ancient' African civilization and not vice versa. They point to the importance of Cush, and the Nubians in upper Egypt in transmitting Benin urban culture north, and black merchants taking 'civilization' across to Greece. Nubian kings conquered Egypt in 712 BC establishing Dynasty XXV. The image of a black Egypt gained much popular interest with Michael Jackson's video 'Remember the Time' (February, 1992). Likewise, Rastafarianism encourages identification with Ethiopian culture. The theory has been propounded by Leonard Jeffries, a very controversial figure, at the City University of New York (*The Times Higher Education Supplement*, No. 1013, 3 April 1992: 11), that the earliest town planners were black Africans, and that the Black Sun people of the southern continents were more creative than the cold, repressed, White Ice people of the northern lands. Quite a variation on the Eurocentric version of climatic determinism. Mumford barely comments on people's colour either way.

Classical zoning

Many of the beliefs which have shaped western, urban reality can be traced to Greece and Rome. Classical cities manifested spatial divisions between male and female realms, which were expressions of associated public/private dichotomies. Because these cultures have had such a profound influence on western civilization it is difficult to separate out the aspects which are specific only to town planning. The social construction of Greek democracy and its embodiment in the πόλις (*polis*, or city state) marginalized women, slaves, and several categories of free male workers. Although the Greek city state is often held up as an ideal town planning prototype, probably less than 10 per cent of the population counted as citizens. The city state of 5,000 citizens, the basis of the folksy, neighbourhood community concept applied in twentieth-century new towns was arguably much larger when all the 'invisible' people were included. Much has been written from a feminist perspective about the influence of the Greek philosophers on modern women's lot (Jagger, 1983). Plato is generally given the greatest blame, for proposing the need for an elite corp which was predominantly male, of the philosopher kings (or guardians), to run society. The guardians were to be kept separate from ordinary people, so they did not become softened or distracted by the trivia of domesticity and family life within the private realm.

Plato apparently had little idea of the practical, daily needs of ordinary people (Okin, 1979) and the activities that kept cities functioning, such as trading – particular malice being reserved for shopkeepers (Plato, 384 BC [1926: 53]). A dislike of trade, and love of sport is amplified within the British education system, and in a classical university education. This influenced the mind-set of many of the founding fathers of modern town planning, who appeared to favour public open space over private. This was epitomized in the importance given to the playing fields: arguably the most sacred, male space. In planning literature, much is made of the importance of Mediterranean city squares, as outdoor living rooms. Constantina (Dina) Vaiou a modern Greek woman town planner gives the example in her doctoral dissertation of the outright opposition a group of women encountered. The police tried to arrest the women when they sought to hold a meeting in an Athens square. She comments, 'we [Greek women] search for the image of a public square and a city belonging to us as well' (Vaiou, 1990: 274). Platzangst (square fear, agoraphobia) was clinically identified in 1873 as a malady brought on by the knock-out scale of modern town planning (Collins and Crasemann Collins, 1965: 157 n186). Woman's private experience of public space may be very different from man's experience of the same space. Yet, paradoxically Vaiou (1992) warns against making too rigid, dualistic, gendered interpretations of urban space.

In ancient Greece, if women stayed within their allotted role and space their lives could be quite reasonable, within the οικος (*oikos*), that is the household. The domestic realm was respected as a separate sphere within which the state did not interfere, for better or worse (Boulding, 1992: I: 227). Woman could have considerable power, provided she accepted her role as the good mother presiding over the private realm of hearth and home and looking after the fruit of the loins (significantly ζώνη = zonē, loins). She must not seek to invade the public space, or challenge the public/private dichotomies which underpinned society. The etymology of the concept of 'zoning' shows the murky roots of this apparently scientific planning principle. Marilyn French points out (1992: 76) that the Hebrew for prostitute, זֹנָה *zonah*, means 'she who goes out of doors' (outside the marital home), that is, she is in the wrong place. It is a specific cognate of *zanah* (Newsom and Ringe, 1992: 197). In Latin, *zonam solvere* means 'to loose the virgin zone', to get married or lose one's virginity; *zina* is Arabic for adultery; and *zenana* is Hindi for women's quarters (Lateef, 1990) (from *zan*, Persian for woman). *Purdah* is part of another etymological string laden with dichotomous images of purity, separation and seclusion, 'within the veil'. In many cultures internal space is graded in importance according to whether it is used by men or women or both, with demarcations between public (visitor) and private (intimate family) spaces (Spain, 1992: chapter 2; Wigley, 1992: 337). Consider the internal zonal concept in French culture of the intimate family space of the *foyer familiale*. *Foyer* means hearth, but it originally meant the focus, or fixed point around which everything (and everyone) is ordered.

Many feminist writers have pointed out the contradiction in the two dominant images of women, as good or bad, angel or whore. The ideal of the sacred virgin, as against the profane whore, was epitomized in the cult of the goddess in the Parthenon (Warner, 1978). Whilst some modern feminists might see the image of a mother goddess as powerful and creative, Freud associated the mother principle with 'stagnation' (reported in Bologh, 1990: 14). Geddes, an early twentieth-century planner, was enamoured of Freud's ideas and these undoubtedly shaped his approach to land-use zoning (Matless, 1992: 466). In spatial terms the good/bad woman dichotomy translates into 'her inside', the angel of the hearth in the private zone, as against 'her outside' (Wigley, 1992: 337). (Compare an article in *Planning Week* (14 October 1993) on planning and prostitution: no space for women planners in this world view.) Levine notes that 'public women' meant prostitutes in the nineteenth century (1987: 147). Boulding (1992: II: 83) states that 'women of the town' had a similar meaning in medieval times. Thus the word 'zoning' contains within it the idea of separation of sacred and profane, especially of male and female (with an emphasis on containing bad woman), and by extension it also carries the meaning of maintaining an ordered, uncontaminated society, through the division of the clean/unclean, *kosher/tref*, pure and diseased. The words 'zoning' and 'sanitary' are etymologically linked (consider *cordon sanitaire* (quarantine zone) and *urbs sana* (the healthy city) (cf. Armstrong, 1993). Thus land-use zoning is a religious and moral exercise. Its alleged practical value, so stressed by town planners is quite secondary and incidental.

Zone zappers

I may be giving too simplistic an impression of women's place in the city, because some women crossed over from the private to public zones. Boulding (1992: I: 227) identifies more than five types of Greek women who were not bound by the constraints of staying in the household, but who could move around in the public zone too, namely intellectual women, older women, poor women workers, slaves, foreign women including traders, and 'εταίρα, the *hetaira* (high-class prostitute). These last blur the simplistic virgin/whore divide, and suggest that class, as well as gender, determined how women were judged, and that definitions of sacred and profane were somewhat different from today. Elite women were allowed to participate in public affairs provided they accepted their honorary position, gave opinions only on what were considered appropriate matters for women, and did not seek to restructure patriarchal society for the benefit of all women. Likewise, in the twentieth century, a certain amount of domestic feminism might be welcomed if demands were limited to women's roles as wives and mothers (cf. Wilson, 1980 in respect of 1940s Britain). Some exceptional women of high birth were allowed to be part of the elite, as philosopher queens, with occasional husband and wife philosopher teams. Such women were allowed to speak out on public matters, and individual personalities might be tolerated as token zone zappers who challenged (but did not actually change) the status quo (cf. Boulding, 1992, I: 231). Elite young men and women might be educated together, in particular for military service, on an equal (male) basis. Consider parallels with planning education today.

Some of these types might be seen as male-sponsored and fulfilling their expected role as patriarchal women, but I am more interested in those who were acting independently, such as women traders, who are potential zone zappers in establishing their own livelihoods and seeking to transcend, if not redefine, public/private dichotomies. Such women might be seen as outcasts in the Graeco-Roman world (Witherington, 1988), but so might some feminists nowadays. Boulding notes the continuing importance in the Mediterranean culture of the tradition of women merchants (1992: II: 278). These were often also travellers escaping the confines of their local community, and thus discovering and spreading new ideas on their journeys. Although the type of the woman trader might on the negative side be conflated with that of the prostitute, the gypsy, and (more honourably originally) the vagabond (Boulding, 1992: II: 80), on the positive side it overlaps with that of the woman scholar and the prophetess. All these are types of the wanderer, who is an essential form of zone zapper. Many positive examples are to be found today travelling around the world in the international academic feminist community, and more specifically among the locales of the women and planning movement.

In the twentieth century, trade in the form of big business (capitalism) becomes a public-realm activity, although still despised by more traditional elite male groups as inferior to public service, which suggests a ranking of entrepreneurial and bureaucratic male fractions (Greed, 1992b: 23). Although women have been excluded from the higher levels, the situation appears somewhat ambigious as to which side of the public/private, male/female dichotomy business should be located in. The private sector provides many different levels and opportunities for newcomers and outsiders compared with more formal, hierarchical governmental structures, thus giving women a potential means of entering the public realm. In the classical world political and military power, rather than capitalism, was the focus of patriarchy. Trade was despised, and so women might actually have had more freedom to be independent business people without condemnation or interference, contributing greatly to the provision of vital consumer goods and services. Women's involvement in trade may partly explain Plato's dislike of traders, who might have included strong, independent women shopkeepers, low-caste freemen, itinerant tradesmen, gypsies (Boulding, 1992: I: 287), and foreigners, all of whom did not readily fit into the either the public or private realms and were not easily controlled. In comparison, in 1993 a draft advice document on gypsy sites from the Department of the Environment made much of the need for clear boundaries demarcated by hedges around the sites where these 'outsiders' established themselves. Women traders, although outsiders, may be seen as the forerunners of bourgeois feminists, or businesswomen today (Greed, 1991: 9). The entrepreneurial woman, along with the philosopher queen, and the femocrat, were to prove the most likely types of women to act as zone zappers in restructuring cities in the past, and are identifiable types in the modern women and planning movement.

Gendered intellectual baggage

Classical urban form may be seen as a mixed blessing. Later Roman towns, especially garrison towns, were built to a standard plan reflecting male/female spatial dichotomies. It is said that every soldier carried a town plan in his rucksack, city-building being an essential means of rapid subjection of newly acquired colonies (Mumford, 1965). In comparison, Bryson (1992: 28) quotes Emmerson on nineteenth-century utopianist town planning: 'not a reading man but has a draft of a new community in his waistcoat pocket'. Nowadays, it may be argued that every male town planner carries around in his head a patriarchal mental map of how a town ought to be planned (cf. Gould and White, 1986; and Nuttgens, 1982 on 'landscape of ideas'). See-sawing from the question of such public plans: at a personal level, garrison towns always bring sexual dangers and controls for women (Eskapa, 1987: 60). But some women achieved an independent role as small traders, their businesses benefiting from the public works and roads programme throughout the Roman Empire, such as Lydia, the seller of purple in the city of Thyatira (Acts, 16: 14; Witherington, 1988: 147).

Although classical architecture appears more feminine in style than that of the modern movement, many of the features have necromantic and erotic meanings. However, unlike modern patriarchal pornography, women as well as men had an input in the development of this visual imagery. The harmony of Greek architecture is achieved by the use of the golden rule of proportion, which although usually seen as a male development, was apparently perfected by the women in the Order of Pythagorus (c. 580–500 BC) (Osen, 1974). The study of mathematics was linked to the understanding of mystical spatial principles and was not purely a male scientific pursuit, but also an expression of women's spiritual dimensions. Centuries of a 'men only' western architectural profession projected a coating of male purity and pristine whiteness onto classical architecture (although many Greek buildings were originally decorated in strong colours). Le Corbusier justified the proportions of the standardized module he used in his mass-production, high-rise housing by implying that his principles were gathered from a study of the Parthenon. But, his module was based on the size of the average male, in spite of the fact that the Parthenon was 'female' in proportion (Pardo, 1965).

Representations of women were sometimes literally incorporated into classical buildings as in the case of caryatides, which outnumber telamones (topless mythological male figures, used as supporting pillars). Caryatides were, incidentally, popular in neo-classical Victorian architecture, appearing 'topless' (but presumably sexless) on such unlikely locations as churches and public libraries as symbols of culture (Sturtevant, 1991: cover photograph). In contrast, modern representations of naked images in public can be seen as threatening by women, reminding them that they are only sex objects and vulnerable, and that in entering public space they are 'asking for it'.

Medieval celestial spheres

The middle ages, in Britain at least, are often presented as the period of vernacular, organic, folksy village settlements. An idyllic image of rustic life was projected back onto this period by some proponents of the garden city movement who were reacting against the horrors of the Industrial Revolution (Darley, 1978). Some of the most limiting conceptualizations of women's place in the home and neighbourhood were inspired by this romantic image. These were often accompanied by reactionary perceptions of women as full-time housewives, in spite of the increase in frequency in the employment of women outside the home in the nineteenth century. Ironically, in medieval times the home/work dichotomy was not necessarily strong, because prior to industrialization many workshops and businesses were under the same roof as the living quarters, with all the family – male and female – working in the family business or on the surrounding land. *And*, unlike today, both men and women were more likely to share domestic tasks too; their home and work roles intermingled (Boulding, 1992: II: 74).

The development of an elite male priesthood and monastic communities in which contamination with women was prevented, enforced the patriarchal nature of western society. Women were seen as 'the devil's gateway', a phrase attributed to Tertullian, *c.* 200 AD. Provided women remained virgins and existed in a separate sphere they might be redeemed. The domestic realm of family life was seen as a second best as a concession to human weakness, and no longer as a respected realm in its own right, complementing the public realm. This is a deterioration from the status given to the realm of the household in the classical world. The separation of some women in convents gave them space to develop their own scholarship and ideas (Warin, 1989). Abbesses such as St Hilda were not just mystics, but more like city managers or town planners, running what were effectively small urban communities and organizing agriculture, drainage schemes and construction. Such women also acted as scientists and physicians (Herzenberg *et al.*, 1991; Phillips, 1990). Women who were married to feudal lords had limited rights under Norman French law. Before the Norman conquest British women had more autonomy over their possessions and lives (Hirschon, 1984). Celtic women had previously had greater legal rights and arguably greater mystical powers (the two often go together). Ancient Welsh law was especially favourable towards women in respect of real property and matrimonial matters (Moody, 1991: 195). With the growth of property ownership over the centuries, the Norman system of English land law which had been created to protect the patriarchal rights of the great dynastic families became, by default, the model for ordinary people's home ownership, the man being seen as the head of the house and its natural owner (Hoggett and Pearl, 1983).

Several imaginary utopias were conceived over the 1,000 years stretching from Augustine's City of God (426 AD) to Thomas More's Utopia (1516). The only mention of women and planning by More is that he suggests women should be occupied 'planning the meals' (p.58 of 1989 edition). There were a few women's utopias, such as Christine de Pizan's work *The Book of the City of Ladies* (1405), to which Bammer (1991) refers in her work on women's utopias, which was more concerned with courtly niceties than radical change. Following the Reformation a new wave of utopian ideas developed, with radical, dissenter and puritan sects such as the Anabaptists, Mennonites, Quakers and Moravians offering a wide range of alternative life styles and community structures. Women often had a powerful role, as equal co-believers. Such sects initially put less emphasis on an exclusively male, separate, priestly hierarchy. Note, John Bunyan (1678) gave as much import-ance to Christiana's journey 'from this world to that which is to come' as to Christian's. Such visionaries were the forerunners of the model community builders of the nineteenth century. Unfortunately on many planning courses religion is merely mentioned as a facilitator of capitalism (Tawney, 1922) and oppression, with Calvin's Geneva (begun in 1541) being picked out as the archetype of urban theocracy. This was a very male example compared with other examples, especially those which were to emerge in North America within a non-conformist utopian tradition (Hayden, 1976).

Urban order

The return to classical ideas in the Renaissance potentially brought back into play the public/private dichotomies associated with Graeco-Roman plan-ning. However, classical planning features were to be used more for aesthetic than functional reasons. The aim was to create fine buildings and squares for the upper classes (male and female), as a stage set to urban life rather than as a reflection of functional gender divisions. City planning on a grand scale often required the clearance of ordinary people's houses, for example to make way for the triumphal route up to St Peter's Square in Rome. Plans were often impractical, with the people being contained out of sight, squashed into apartment blocks, between wide public boulevards, which were designed to give the right perspective vista, and to enable troop movement to quell urban unrest (Sutcliffe, 1970). From approximately the sixteenth to nineteenth century mainland Europe was in the grip of 'Grand Manner' planning. The creation of total urban reality was epitomized in the activities of the Sun King, Louis XIV of France (1638–1715). His patronage of art, architecture, music, dance, literature and town planning was intended, to create an illusion of civilized culture. Visually the knockout scale of such European planning was impressive, compared with the more subdued Georgian style in Britain. The continental style of city living was un-doubtedly more urban, sophisticated and higher rise. It has been suggested that after the virtual unification of the country under William the Conqueror there was no need for internal defences in England and therefore urban form naturally tended towards a more garden-city, suburban dispersed style with gardens and orchards. In contrast, in Europe, city states were still fighting it out into the nineteenth century, and it was dangerous to lower one's defences in order to extend the city walls. People built upwards rather than outwards.

This created a much higher density, and a more communal form of living, often with different social classes living like a layer cake above or beneath each other in the same block (Sutcliffe, 1974). Less space for domestic chores led to a stronger tradition of public provision of laundry services, cafés, childcare and leisure outside the home, so there is no exact equivalent of the trapped suburban housewife of Anglo-American literature, but other problems instead (Strauch and Wirthwein, 1989).

Georgian architecture in Britain, although appearing more homely, with its patchwork of town squares, embodied public class distinctions. Space for family and servants was horizontally divided within town houses; servants (consisting of both women and some men, and therefore not an exclusively gendered, internal division, McBride, 1976) occupied the attic rooms and worked in the basement kitchens. Rateable value of residential properties was fixed according to the floorspace, on the basis of six rates (Clarke, 1992: 112). Developers tended to build up to the maximum allowed for a particular rate, resulting in house types visibly reflecting, even enforcing, social order and income divisions. Possibly, class divisions were more important than gender divisions in such areas in ordering urban form (Davidoff and Hall, 1983). Indeed, some women had considerable influence on development. They acted as leaders of fashionable society in deciding which were going to be the 'good areas'. They established intellectual circles, and increased women's presence in public: redefining the drawing room as a public space (Clarke, 1992: 18). Women still act as urban trend-setters and catalysts of gentrification (Hinchcliffe, 1988). Bath developed as a spa town because Queen Anne 'took the waters' there in the early 1700s. Some women were involved in development such as the Dowager Duchess of Bedford, who in 1774 began building on a 112-acre site in Bloomsbury (Ashworth, 1968: 36) which was later to become the habitat of the Bloomsbury Group in the early decades of the twentieth century. Before the formalization in the nineteenth century of activities such as surveying, architecture and planning into exclusively male activities called 'professions', women were actively involved in professional practice, some running building firms, such as Maria Savill (Greed, 1991: 46). Even so-called 'fallen women' had an influence on town planning: the main axis of Nash's Regency plan for London purposely divided Soho from more respectable adjacent areas. Summerson, significantly, called this division 'biological' (1976: 186).

REVERTING TO TYPE

Thus the seeds of what is wrong with cities were planted generations ago. Assumptions about the need for the gendered organization of space to express public/private dichotomies might be so deeply embedded in people's minds, and in western civilization as whole, that demolition, bombing, or the imposition of a revolutionary plans will not eradicate the tendency to revert to a particular type of urban form. Indeed, urban conservation policy may be perpetuating gendered patterns. Much of this happens automatically as people appear to be locked into the assumptions and 'patterns of the past', as Patsy Healey, Professor of Town Planning at Newcastle, has highlighted (in discussion arising from a joint presentation on the future of women and planning to Scottish women planners (*The Scottish Planner* No.26: 10, April 1992, and Greed, 1992d)). It is a two-way process. The city is the product of the reproduction over space of social relations but, once built, the physical structure can, in turn, feed back its influence onto its inhabitants, by acting as a constraint on the nature of future societies living in that city because of the restrictions of its layout, street pattern, design and subculture.

6

THE NINETEENTH CENTURY

INDUSTRIALIZATION OR MASCULINIZATION

Just a few mad women?

I do not intend to give a comprehensive historical account, but to highlight the influence of beliefs in public/private dichotomies on the conceptualization of town planning. First, to provide a context, I will consider how the town planning agenda prioritized issues related to the (male) working-class and industrialization. These concerns fell within the public side of urban affairs, but had implications for the private sphere of domestic life. This is the period when domestic ideology was promoted, and the first-wave of feminism developed, as women, especially middle-class women (cf. Levine, 1987) faced increased social and spatial restrictions. Second, I will outline the mainstream town planning movement, highlighting its gender bias. Those unfamiliar with the period are recommended to consult more detailed accounts (such as Ashworth, 1968; Cherry, 1974, 1981; and see Appendix II), but with caution. Booth (1986) in his article on textbooks most frequently cited for this period lists no books by (or presumably about) women. However, redressing the balance somewhat, Gordon Cherry (1991) in an article on recent developments in planning history does mention women and planning. Third, I will rerun the spool to identify women's contribution to the development of town planning, highlighting how women as planners (in various manifestations) acted as zone zappers seeking to break down public/private dichotomies.

Like Dale Spender (1982: 4), as a student I accepted the received tradition which held that a few unbalanced women chained themselves to railings in an attempt to get the vote, when I was to discover there was a first-wave urban feminist movement which was as productive in ideas and literature as the early malestream town planning movement. Women were ahead of men in town planning matters a hundred years ago, and the male domination of town planning during the twentieth century is nothing but a temporary intermission in the progress of real town planning. Far from the present women and planning movement being a sign of progress, it might be seen as but a pale shadow of what went before, narrowly confined within patriarchal govern-mental structures, and uneasily allied with peculiarly male political ideologies (Greed, 1992a). At the turn of the last century town planning was at its zenith as a visionary social movement and had not become formalized into a bureaucratic, patriarchal profession embedded within local government structures which developed in the twentieth century. We are playing a long game here. Like horses in a race across the course of the last two centuries (1800–2000), some themes appear to dominate and then fall back, and incompatible running partners may run alongside (to block others passing). The early front runners in town planning were communitarianism, first-wave feminism, utopianism, evangelicalism, temperance, the arts and crafts move-ment, colonialism, and eugenics; to be replaced by municipalism, socialism, scientific objectivism, urban managerialism, neo-Marxism, second-wave feminism, enterprise-culture planning, environmentalism and urban design-ism. Therefore, the chapter concludes with a consideration of some of those 'other' elements which were so important to first-wave urban feminism.

The making of the working-class

Much is made in the contextualizing literature of town planning of the importance of the working-class as the group who suffered, and for whom it was all 'for'. But this class often appears to be seen as part of the public realm, not as being made up of 'private' individuals: class conflict being depicted in abstract terms as the result of impersonal forces, as separate from how people felt, and how they treated each other at the inter-personal level. The working-class appears as chiefly male. Whether women are included as non-gendered workers, is questionable in view of the masculine nature of the subjects of which the discourse is composed. Exaggeration of previously uncertain divisions between male and female occupations and spaces was achieved through policy-making that was ostensibly concerned with solving the problems caused by the Industrial Revolution. Town planning, especially zoning of industry and the creation of separate, residential neighbourhoods, was one means of enforcing divisions spatially between male and female, and of preventing women combining work inside and outside the home in their daily lives. A rich mythology revolves around the adventures of the male working-class, often depicted as the heroic, muscular, half-clad, sweaty manual worker, an iconography which made many women feel uneasy, and reflected other subterranean, male homo-erotic streams within the labour movement (Tickner, 1987: 180). The title of E.P. Thompson's classic book *The Making of the English Working Class* (1963) takes on a new meaning in terms of 'creating realities' (cf. Berger and Luckman, 1972). Many of the venerated texts were written by Oxbridge men, such as Briggs (1968), Ashworth (1968), E.P. Thompson (1963), and Trevelyan (1978). Ostensibly concerned with the working-class, much of the limelight seemed to centre on the importance of dead, middle-class men with beards (see p.78 on DWEMS), that is, philosopher kings and 'geniuses' (such as Shaw, Geddes, and Marx). These figures appeared to speak 'for' these oppressed men, not only in the history of planning but also in architecture, economics, socialism and sociology. Looking at some of the history books in the light of feminist scholarship, one wonders how the authors got away with presenting such a partial view, in which it was essential to pretend that there had never been a feminist movement (Delamont, 1992). Teather (1991b) implies that many contemporary planning documents are nothing but 'value-laden fiction', a phrase which also aptly describes much patriarchal work on the nineteenth century.

Large numbers of women were involved in industrial activities outside the home and thus were workers in the patriarchal sense. Women were also workers within their own homes, and in the homes of others. Tannahill (1989: 347) states that there were 1,204,477 domestic servants in 1871 (Roberts, 1991: chapter 2, puts it higher). In the nineteenth century this was the single largest employment category for women and the second largest overall, but these people were not seen as a significant class of workers. However, the relative marginalization of domestic servants in the conceptualization of the working-class, cannot be explained by the fact that they were female. There were male servants too, but it may be argued that an increasing feminization, and marginalization, of domestic labour was taking place (McBride, 1976). The remaining men took on supervisory or superior posts as footmen and butlers (cf. Davidoff and Hall, 1987: 391). This created complex, nested public/private; male/female hierarchies within the domestic workforce, and the private spaces of the home. But, the philosopher kings chose to see domestic labour as simply a non-work activity undertaken chiefly by females.

Hygiene and filth

Although women were often excluded in the conceptualization of theories of class they were conveniently included back into the human race when needing someone to blame for the problems of nineteenth-century society. But, women were judged and treated differently according to their class position: dichotomies between working/middle-class, or unacceptable/acceptable women enforced divisions among women. If working-class women are mentioned in the mainstream historical account, they are portrayed as causing the problem of overpopulation because of low morals (or inadequate eugenic teaching, McLaren, 1978; Mazumdar, 1991). Working-class women, especially those in the slums, were often blamed when it came to matters of disease and social hygiene (Richardson, 1876; Jones, 1986). Ostensibly this was because of their failure as housekeepers and mothers, but such judgments also reflect deeper beliefs about the dirtiness of women as polluters of mankind. Peter Hall (1989: 27) explains that state intervention in the built environment accelerated once it was realized that cholera was a water-borne disease which could affect everyone. A single polluted pump in the Soho district in 1854 (an area associated with sexual pollution: Summerson, 1978: 186) was identified as a source of pollution of the capital's water supply: physical and moral impurity were conflated. The Contagious Diseases Acts of 1864, 1866, and 1869, to control venereal disease in garrison towns, were applied only to women (Eskapa, 1987: 61) and only to 'common' women at that (Levine, 1987: 145). Soldiers were not tested, suggesting that pollution was believed to be only one-way: from female to male.

If middle- or upper-class women are mentioned, they are seen as sexless angels on pedestals; or paradoxically as neurotic and contributing to the breakdown of society (as in Durkheim's work on suicide, 1897). Woman's presumed selfishness, that is, her preoccupation with the private domestic realm rather than the public common good, might be cited as the 'problem' (Callaway, 1987). This removed the blame from patriarchy. Seldom does the literature reflect that women have needs of their own. Women, both as wives and servants, experienced tremendous housework problems in the nineteenth century that increased as middle-class houses became larger and more ornate. But it was considered unacceptable to control the design of middle-class housing, although the sanitation often left much to be desired (Rubenstein, 1981). Such housing was also very cold, as in the case of the architect Lutyens' large townhouse in Bloomsbury Square as described by his daughter (Lutyens, 1980: chapter 4).

In contrast, the control of working-class housing was fair game. In 1842 Chadwick's report on sanitary conditions resulted in the 1848 Public Health Act, which introduced minimum physical housing standards. Although passed in the name of public health, some would argue that such measures, (and the controls on common lodging houses) were concerned with sexual morality – overcrowding being a euphemism for incest, prostitution and adultery – although incest did not become illegal until 1908 (Wilson, 1991: 35). The concern was probably a eugenic one, motivated more by a pre-occupation with the dangers of inbreeding, than with the human cost of sexual abuse. There followed a series of acts including the 1875 Public Health Act. The 1890 Housing of the Working Classes Act created the first 'council houses'. The public health, housing, and town planning movements have continued to control the working-classes ever since (Cockburn, 1977). Housing policy was subsequently used, alongside town planning, to control ethnic minorities in the twentieth century (Smith, 1989). Likewise the domestic science movement was seen as a means of modernizing domestic labour and reducing women's workload, but became associated with the control of working-class schoolgirls and thus housewives (Attar, 1990). Domestic science was linked to the women's town planning movement, as in the work of Ellen Swallow Richards, who was the first woman science graduate of MIT (Hayden, 1981: 150–9).

THE DEVELOPMENT OF MALESTREAM PLANNING

Men's model communities

New Lanark was one of the first model industrial communities in which town planning principles were practised in Britain (Bell and Bell, 1972). It was developed by David Dale as a planned community of 2,000 inhabitants, but Robert Owen took it over in 1800, when he married Dale's daughter (and fortune). Colin and Rose Bell comment that she was 'as anxious as any girl to marry' (1972: 244). Over the next 25 years Owen developed ideas on education, health, housing, trade unionism, callisthenics (early aerobics) and birth control. He founded the Institution for the Formation of Character to shape his workers' minds. Whilst New Lanark gets into all the planning history books, one senses that there has been a filtering out of those communities based on other, less acceptable ideologies, especially those religious and feminist ones which challenged gender roles (Pearson, 1988; Hayden, 1976), and a focusing on those based on meeting the needs of the Industrial Revolution, and its presumably male workforce.

Many of the model factory towns, including New Lanark, were not based upon having a purely male workforce with women at home as full time homemakers, although one might get this impression retrospectively from the mainstream literature. In New Lanark, the women worked as well as the men. One of the functions of the nurseries set up by Owen was to free women workers from childcare as well as to enable him to shape the hearts and minds of his future citizens. From a feminist perspective New Lanark may be seen as paternalistic and patriarchal in character, because although welfare was provided, the people themselves (especially the women) were not in control of its administration. New Lanark was not a democratically created community run by its residents, but a top-down attempt to mould the lives of a tied working-class. It is hard to judge how happy the residents were with their lot because, as with wives in marriage, if they left they would lose their homes, and face an uncertain future. Distrust seems to have been reserved for the women as homemakers: Owen inspected the dwellings for bedbugs. It seems to me this reflects the beginning of a trend, manifest in the Welfare State today, of the public realm invading and seeking to take over the private, domestic realm, taking away the autonomy (albeit limited) that women as household managers (and working-class men as heads of household) might have had when the domestic sphere was respected as separate from state control. In American, Owenite communities women acquired a further burden of collectivized community housework, under the supervision of men who were not even their husbands (Kolmerten, 1990). In contrast, Owen's daughter, in the classic dutiful role of daughter-of-famous-man, was trusted to have some independence, in order to undertake speaking tours about her father's social experiment (RTPI, 1986: 30).

Some women colleagues disagree and say that Owen was really quite a feminist, and the model communities movement was not part of a conspiracy to subjugate women or the working-classes. It would appear that Owen was cognizant of early feminist ideas, and familiar with wider communitarian ideas (Owen, 1849) such as those of Fourier (Goret, 1974), who also favoured the apartment block to the individual house and was part of a wider French circle of socialists, feminists and free thinkers.

The activities of the Bradford-Halifax school of philanthropic factory owners is usually seen as the next significant stage, from the 1840s onwards. Titus Salt (1803–76) founded Saltaire, a small scheme of 800 houses (and named some of the streets after his female relatives). He is better known for being the first industrialist to introduce the official tea break and for seeking to free his workers by means of encouraging the male workers to buy their houses on mortgages, through what was to become the Halifax Building Society (Ashworth, 1968: 131). From the beginning building societies have been patriarchal in terms of both lending and staffing policies (Crompton and Sanderson, 1990: chapter 6.) Although Salt provided several communal facilities including wash-houses, his provision was partial – domestic labour was not collectivized – and he objected to women putting their washing out over the back of the houses (Nuttgens, 1972: 89), and used to gallop down the backs cutting the lines with his sabre.

Garden cities

The model communities movement culminated in the garden-city movement and the work of Ebenezer Howard. He was a self-made man and visionary (the maverick alpha-male) who began life as a shorthand typist working in the law courts. In those days typewriters were high technology, comparable to computers today, and typing was men's work (Delgado, 1979). Briefly, in 1898 Howard proposed garden cities of around 30,000 people divided into neighbourhoods of 5,000 that would be focused around a community centre and school, with clear land-use zoning and a green belt around the city. A series of garden cities was to form a ring around existing conurbations and act as counter-magnets, attracting people to decentralize, thus creating balance. Howard stated that town and country must be 'married', and 'out of this joyous union would spring a new hope, a new life and new civilization' (quoted in Darley, 1978: 181). Within the garden city overall densities would be relatively low, with lots of grass, trees, gardens and (hopefully) sunshine, thus diluting the squalor and overcrowding of urban life. His ideas embodied many of what were to become the main features of twentieth-century British town planning (Cherry, 1981; Hall, 1989).

Howard is one of the good guys compared with some of the patriarchal proponents of town planning who were to emerge in the twentieth century.

Like Mumford (see p.76) he was somewhat of an outsider in his day. However he was later recreated as the grandfather of British patriarchal town planning. Certain aspects of his ideas were stressed, even exaggerated, while his more radical ideas were forgotten in order to legitimate, retrospectively, the particular form of town planning which developed in his name in the twentieth century. Howard was conscious of many of the dichotomies which split British society, over and above those which generated the urban/rural divide which his garden city sought to reconcile. He envisaged the ideal city being run on communitarian principles, as he put it, driving a wedge between the vested interests associated with capital and land, transcending present economic and political constraints and setting a new direction for progress (Howard, 1898: 147). In the twentieth century, emphasis was put upon the spatial aspects of his proposals in isolation from aspatial ideas on land ownership and urban economics.

Howard was a zone zapper himself; the separation between home and workplace he proposed was really quite small-scale, and mainly based on walking distances. But his ideas were reinterpreted to justify large-scale zoning, especially the division between residential and employment zones. He envisaged that people would travel by public transport, and he was a strong advocate of the bicycle to make all his city accessible to everyone, so he incorporated cycle works in his original garden city plan. The bicycle was a new invention, much favoured by 'liberated women'. He acknowledged the problem of domestic labour, and planned one neighbourhood at Letchworth, called Homesgarth, on co-operative housekeeping principles (Miller, 1991) with shared, centralized kitchen facilities. This was but a faint reflection of feminist town planning principles, and like many middle-class men he was probably more concerned with dealing with the servant problem than with restructuring home/work, male/female dichotomies (Ravetz, 1989: 192). The Howards lived for a short while at Homesgarth with Mrs Howard, doing the 'co-operating', an arrangement which apparently caused considerable friction in the family. Co-operative housekeeping ideas were edited out of the historical record. Hayden notes (1981: 336, n7) that Purdom, the planning historian, had written at length on this aspect in the 1913 edition of his book on the garden city movement, but reduced it to two pages in 1925, and three paragraphs in the 1949 edition. This is a classic example of the process of rendering feminist ideas invisible.

There is a tendency in the telling of 'his'tory to focus on a key male name which then blots out the contribution of others, especially women. Howard was not a solitary genius, but part of a group of men and women pursuing similar ideas. Bournville, the factory town built on garden city lines by George Cadbury, had already been started in the late 1870s, and Cadbury established a chair of town planning at Birmingham University (Gardiner, 1923). Although the Bournville scheme embodied many advanced planning ideas it reflected gendered priorities. Cadbury's plan allowed for a large 'men's recreation ground' in the town centre, and a much smaller 'girls' recreation ground' squashed in alongside (Fischman, 1977: Figure 2). But, Cadbury was chivalrous in requiring women workers to be let out of the factory each day before the men to avoid unnecessary jostling between the sexes. Port Sunlight near Liverpool was built from 1888 by Lever the soap manufacturer (Cherry, 1981; Ashworth, 1968). The houses, although surrounded by extensive open plan grass areas, were mainly grouped in blocks without private back gardens. Much to the annoyance of Lever, housewives used to put their washing on the fences around the houses as no washing lines were 'planned' (p.16, n2). The designs of architects Raymond Unwin, Parker, and Lutyens (Service, 1977) were employed in the planning of various garden cities and garden suburbs.

Howard was an internationalist, a keen advocate of Esperanto, and vege-tarianism. He died in 1928. Fischman (1977: 82), in a section entitled 'Beyond the Grave', refers to a seance in 1926 at which Howard apparently received a message from his first wife, 'you have accomplished more than you know', (from the Howard papers letter to G.G. André (a spiritualist) 18 October 1926, Folio 25). He had taken up spiritism in 1904 when his first wife died. He remarried in 1907 and his second wife, who died in 1941, was herself involved in planning circles. Howard was open to a range of 'other' philosophies, and was far more of a Renaissance man than the narrow bureaucratic personalities who were later to practise town planning in his name. Some of the other 'great men' of the early town planning movement were involved in the spiritual movements of Edwardian times. The historical record has cleaned them up and presented them to us as mono-dimensional modern men. Edwin Lutyens the architect was caught up in theosophy (Lutyens, 1980) through his wife, who was a follower of Mrs Besant. Besant was President of the Indian National Congress. (Incidentally Sonia Gandhi (an Italian) in 1991 was the only other European woman to achieve this post.) Lutyens had strong links with India, designing the Governor's Palace in New Delhi. However, it was low-caste Indian women who built the Palace, moving every stone by hand, and who still do much of the unskilled work on building sites (Carter, 1991). Thus European ethnicity was a key factor along with class, in determining the woman who was given a policy-making role. The Indian women who were *in* the construction industry were ignored. Besant was a supporter of eugenics, and was a Malthusian (as was Shaw, and Howard manifested these tendencies to some extent, Home, 1993: 15; Voigt, 1989).

WOMEN'S PLANNING HERITAGE

Overlaps or plagiarisms?

Women were among the first to support, even design, the garden city concept. Angela Burdett-Coutts, a philanthropic millionairess who had carried out housing schemes in Bethnall Green for the working-classes, in 1865 built Holly Village in Highgate incorporating many village features which anticipate the garden city concept (Darley, 1978: 72). The Marchioness of Waterford made picturesque improvements to her estate by model village building at Curraghmore, Ireland (Darley, 1978: 99). Bedford Park in West London, commenced in 1877, although built by a speculative builder, Jonathan Thomas Carr, anticipated elements of the garden city, and attracted a range of prominent intellectuals as residents. Discussion on women's rights and suffrage, with the local community association the Bedford Club, was a feature of community life in which suffragettes such as Lydia Becker and Florence Pomeroy participated (Bolsterli, 1977: chapter 4). Henrietta Barnett continued the garden city tradition with the Hampstead Garden suburb, begun in 1907. Lutyens was of the opinion that she was 'a nice woman but . . . has no ideas much beyond a window box full of geraniums, calceolarias and lobelias, over which you see a goose on a green' (quoted in Darley, 1978: 180). Ironically, whilst women planners had very practical planning ideas, male-stream town planning espoused sentimentalized images of the private domestic realm. However, one cannot be uncritical of the women's ventures into town planning, as by the standards of today they appear strongly, but inevitably, 'classed' in nature.

Inspiration

Whilst women's approach to town planning was shaped by the need to deal with daily, practical urban issues, not least of which was the problem of domestic labour, women also drew on a wide range of literary, artistic and religious sources in 'imagineering' the future shape of cities. The roots of the women and planning movement seem to be much more creative and qualitative in nature than men's inspiration which, in spite of the occasional visionary, was often based upon technical public health criteria, engineering standards, physical rules and laws, and meeting the needs of industry. Howard's garden city was a most 'romantic' version of male town planning, almost feminine in comparison with some of the grim technological solutions to emerge later. But then, a genius must be more feminine in his creativity than women are themselves (Battersby, 1989). Kate Greenaway was also a proponent of the garden city approach, and had a house in Hampstead Garden suburb designed by Norman Shaw according to her directions. She is better known for her illustrations of children in village settings (Ward, 1978b). Her work was relegated to 'arts and crafts' rather than 'town planning', but she was just as much an imagineer, creating reality in her drawings. These were copied in crayoning books, which have socialized generations of children to what village life ought to be like.

The early women and planning movement was part of a wider philosophical and literary feminist upsurge. In the previous century the foundations of the first-wave of feminism had been laid through the work of women such as Mary Wollstonecraft (1792), who herself was part of a wider literary circle of poets, revolutionary thinkers and socialists (Showalter, 1982). Women were critics of existing society and 'imagineers' of alternative societies. They were likely to express their ideas in literature, because they had limited access to the public realm of politics, philosophy and the built environment professions, with some exceptions such as Angela Burdett Coutts, Henrietta Barnett, and the Marchioness of Waterford. Mary Shelley, daughter of Mary Wollstone-craft, created Frankenstein in 1818, possibly as a warning against a future man-made world in which man's science would capture the secret of life, but could go dreadfully wrong. Literary, romantic visions of reality feed back into concrete reality when women have the opportunity to act as developers, and vice versa. Jane Austen had earlier passed social comment on planning of the landscape in *Mansfield Park* (1814). The woman landscape gardener, as epitomized in Gertrude Jekyll (Massingham, 1984), is a continuing, acceptable type of woman planner, possibly because the profession is seen as more feminine than urban planning. Gillian Darley the architectural historian (1978: 98–9) explains how George Eliot's heroine in *Middlemarch* (1871), Dorothea, was based on the Marchioness of Waterford. The ill-fated Marie Antoinette had enjoyed creating rural realities, such as were already expressed in French art at the time, playing at being a *bergère* retreating to her *chaumière* on the royal estate. Some women novelists were linked to model community coteries in North America, Louisa May Alcott, of *Little Women* fame (1868) spent her childhood in more than one model community, because her father was an enthusiast. That would have been just the sort of childhood connection which would have made me feel more at home as a new planning student, to balance all the boyish references to Meccano and train sets in lectures – and Nintendo nowadays.

Utopianism

Urban feminism was linked to utopianism (Goodwin, 1978). Marx and Engels caricatured communitarian socialism as utopian and emphasized strategies for organizing industrial workers (who were seen as male), thus losing sight of half the human race. Hayden comments (1981: 7) that many men used feminism as a derogatory term to criticize 'political deviation' and 'dis-loyalty', it being seen as a means of preventing women identifying with the 'revolution' as members of the proletariat (Taylor, 1984). I identified a similar problem in former socialist states on p.58. Hayden comments that feminists and Marxists each had a piece of the truth about production and repro-duction, and she sees early urban feminists seeking to combine both parts of the male/female, public/private dichotomy in the design of model com-munities in America as part of a wider utopian movement. Utopianism was a potent social movement, which inspired its adherents with almost religious zeal, to overcome all obstacles, in order to achieve material goals like building new towns or colonizing the Mid-West (Manuel, 1979). Utopianism was also seen as a destablizing movement by the establishment, which often linked it to sexual subversion, presumably because traditional gender roles within the family might be questioned (Kumar, 1991: 86). Modern town planning was to offer neither revolution nor utopia, but reform, which rapidly deteriorated into control to facilitate the further growth of the very factors the early movements wanted to replace, namely capitalism and patriarchy (McMahon, 1985) (cf. Hardy, 1991a, 1991b).

Throughout the nineteenth century, writing utopias was a major craze in Britain and America, with emphasis on more spiritual ones at the beginning and more scientific ones at the end. Both the first-wave of the women and planning movement, and the malestream town planning movement produced vast quantities of utopian and futuristic literature, based on a wide range of ideological and political perspectives. The literature ranged from serious novels to complete blueprints for the transformation of society accompanied by land use plans. Examples include Bellamy's *Looking Backward: 2000–1887* (1888) and Charlotte Perkins Gilman's *Herland* (1915). (Gilman also wrote on town planning issues (1921).) Ebenezer Howard's 'Tomorrow' (1898) is another example. According to Parrington, between 1884 and 1900 around forty-eight utopian romances were written, an output which to a degree compares with that of science fiction today (Greenwalt, 1955: 19, n19, quoting Parrington, 1930; Kessler, 1984). But Albinski (1988: 45), writing on women's utopias in British and American fiction, says there were 190 women's utopian works published at the zenith in 1890–9 alone, 70 of these in Britain. She notes that male writers (such as Bellamy) usually included future metropolitan utopias, whereas women favoured localized, slow changes occurring in contemporary urban situations rather than in future society. This point is developed further by Elizabeth Russell (1983), whose research covers 1792–1937.

Nowadays some feminist visionary works might be written as science fiction (Bammer, 1991), as a means of creating alternatives, or to warn against current trends such as genetic engineering and patriarchal, psychiatric control by drugs. Some also blur the present/future dichotomy so prevalent in scientific thinking in general, and planning in particular (Piercy, 1979). Women also combine elements from the past and the future in swords-and-sorcery sci-fi (Armitt, 1991), using a mixture of space-age and Celtic Mabinogion imagery (McCaffrey, 1990). Women also like to go 'where no man has gone before' and visit in their imaginations other worlds (LeGuin, 1969). It may be argued that women tend to go for inner space whereas men go for colder external worlds as their 'other'. But this may be a sexist stereotype, although women's outer space stories may be more likely to be located underground inside a planet (cf. Vonarburg, 1990). Significantly, from a class perspective, women (as well as men) in positions of authority can be baddies in such utopias which often seem more like dystopias in practice (Atwood, 1987). For some women the creation of cities had to be an internal visionary exercise, expressed in literature, because in the past women had no power to build their dreams (Squier, 1984; Sizemore, 1989). I find it odd that, in spite of the access women nowadays have to the built environment professions to create realities in bricks and mortar, many still prefer the humanities and arts, as if they can still 'only' write. This is not to discount the blocks and barriers women encounter when they try to plan, and the struggles of writing. It begs the question of whether the word processor is mightier than the cement mixer as a means of creating realities and imposing change on society, for, as has been stated, many of the 'great men planners' built very little but wrote a great deal. The city can provide strong imagery to express a variety of aspirations, across a wide range of women's literature, for example in Anaïs Nin's work (1959); who, see-sawing, paradoxically suffered from oppressive patriarchal relationships at the personal level (Millet, 1970: 312n).

Think-tanks

In looking at the development of town planning across the nineteenth century and into the early twentieth century, one must acknowledge the key role of relatively small groups of people in creating the agenda: a self-appointed Plato's elite of philosopher kings and queens. These were not only interested in town planning, but in a range of other interconnecting reforming and philosophical movements, with certain women strongly represented within this network. The Fabians (from the 1880s) and the Bloomsbury Group (from the early twentieth century) acted as key agents in generating social change. Members of these groups were connected to many other like-minded people in the provinces, including slavery abolitionists, quakers (as in Bristol), free-thinkers, socialists, feminists, vegetarians, business people, and radical non-conformists. These significant groups included (at various times) philanthropists such as the quakers George Cadbury and William Lever, who linked across to Ebenezer Howard, Eleanor Marx, Havelock Ellis, George Bernard Shaw, Olive Schreiner (1911), Octavia Hill (Darley, 1990), the Webbs (MacKenzie, 1986), Charles Booth, the social researcher (Booth, 1889), William Booth (1890), the Salvation Army leader (who favoured women in leadership roles), Gertrude Jekyll, Annie Besant, Arnold Toynbee, and Madame Blavatsky. Victorian and Edwardian intellectuals who were publicly linked were often also connected (Brandon, 1991) in a range of private liaisons. For example, Marie Stopes, the birth control pioneer, lived with Havelock Ellis who in turn had been linked to Olive Schreiner.

In addition to feminist and socialist issues, one is suprised by the interest shown in the paranormal and spiritism. The range of interests, and the way they were blended and juxtaposed, to make up such a rich intellectual soup, contrast with the crippling intellectual dichotomies which emerged in the twentieth century, separating scientific and spiritual, and social from spatial. Spender reflects on why so many of the early feminist women took up theosophy and spiritism (1985). Alex Owen argues that women did so in order to influence events because personal and property law at that time denied them any separate existence. If women were married they could not sue or be sued, or own property in their own right (Owen, 1989). Many nineteenth-century women and men town planners were inspired by a strongly religious and moral motivation. Many reformers saw the working-class as immoral and in need of 'saving', and their world views were informed by good/evil, God/Devil dichotomies. Both feminism and town planning were strongly linked to temperance, and evangelicalism. Activists campaigned against working-class men spending all their wages on drink and going home to beat up their wives (Gilman, 1915, 1921; Banks, 1981; Boyd, 1982). Nowadays, the D.o.E declares 'a planning application, for say an amusement centre, should be considered on its land use merits, and not on the basis of issues which are immaterial to those considerations, such as moral grounds' (1991: para 29). First-wave feminist planners would not be able to comprehend this moral/amoral dichotomy, any more than modern critics of planning can understand how one can divide the social from the physical aspects of land use. The women's suffrage movement demanded a single moral standard for both sexes, expressed in one of their slogans, 'votes for women and chastity for men' (Tickner, 1987: 223). The modern manifestation of planning, and of feminism, has often been linked with socialist and materialist (and even libertarian) attitudes. Second-wave feminist planners may have difficulty in understanding why their first-wave counterparts went about things the way they did, and the value systems which motivated them. This may lead to parts of the historical record being rejected by modern women town planners as not being with the ambit of planning or feminism, when the reverse is true.

Town planning was an important issue on the agenda alongside housing, women's emancipation, education, eugenics, and political reform. In spite of the undoubted scope of interests manifested, the type of people who participated were a fairly narrow group in terms of class, and so their knowledge of and concern for the working-class was inevitably somewhat limited. Admittedly their number included some individuals from working-class backgrounds and those who had worked in the slums. The women involved seemed to be above mere mortals. The diaries of Beatrice Webb (MacKenzie, 1986) reflect how busy and influential upper-class women were. Webb played a major part in the founding of London University and advised politicians on parliamentary issues, albeit without the vote herself. She was involved in the garden city movement and hundreds of other good causes in a constant round of activity. No wonder neither Beatrice Webb nor Octavia Hill had much enthusiasm (initially) for the suffrage movement. After all, democracy is for the masses not the philosopher kings and queens. Virginia Woolf said, wisely, she would rather have money than the vote (Woolf, 1929). Such women scarely needed access to education themselves as they drew on an intellectually rich home culture. Education was for the working-classes and parvenues. Nowadays, with the introduction of student loans, which may be seen as negative dowries for women, we may yet see a return to the university of the home. We may be at the beginning of the end of the modern university movement which has been going strong for over a century now. It seems to me that the higher the class of certain women the less they were affected by public/private dichotomies. As we shall see, the lower the class the stronger the division, and the more likely the public authorities were to interfere in people's private lives.

Women's model communities

In the USA, there was another group of founding mothers of town planning, the Harvard Square Group, who acted as zone zappers creating ideas as radical philosopher queens. They also sought to build experimental towns, taking on the role of traders who entered the property business, that is as entrepreneurial developers. Amongst them were Melisia Fay Peirce, Catherine (Beecher) Stowe, Charlotte Perkins Gilman, Ellen Swallow Richards, and Alice Constantine Austin. I do not intend to describe their work here, but refer the reader to Hayden's comprehensive account (1981). Some readers might be thinking, 'well that's America: not Hayden again', but malestream British town planning has always been unashamedly linked to America. Hall distinguishes the Anglo-American school of planning as a grouping distinct from the European (continental) school (Hall, 1989: chapter 3). As Cherry points out (1981: 41) the first president of the TPI (Town Planning Institute), Thomas Adams, 'gave to each side of the Atlantic a son who became the leader of a generation of planners'. James became president of the TPI in 1948 and Fred was head of MIT's city planning faculty and president of the American Institute of Planners. This is but one of the many examples of the fraternal and paternal linkages and dynasties one finds running through the world of town planning, and prevalent in the world of public service, that marginalize women and other newcomers. There were examples in Britain too, albeit on a smaller scale. Several schemes are still standing, although their creators are forgotten (Pearson, 1988; Greed, 1993b).

Many experimental towns were built in North America for a variety of religious, political and ideological reasons, including ones with a substantial urban feminist input. Everything we barely dare dream of, they had tried nearly a hundred years ago: new towns, kitchenless houses, co-operative housekeeping, reconceptualization of land-use zoning, public transport systems, and 24-hour childcare. Many of the principles of women's alternative approach to town planning culminate in the design of Llano del Rio built by Alice Constance Austin in California in 1916, which I see as a counterpart in importance (and mixed success) to Howard's Letchworth Garden City in

Britain. Her development was based on the principle of co-operative house-keeping to free women from domestic chores, by centralizing the provision of laundry, cooking, childcare, and cleaning services, a common feature of many feminist schemes of the time. Thus most of the dwellings were kitchenless houses. Personally I am not so keen on this aspect, as people like the ability to make a pot of tea when they want to, although ready-made meals sound wonderful. Schemes were proposed in which the town would be based around a socialized, centralized food production facility, with electric conveyor belts sending the hot food around the town, with people putting their dirty dishes back for washing up to take place back at the central facility (Hayden, 1981).

The borderline between centralized support for domestic tasks and child-care, and the need for individual freedom and privacy is always a tenuous issue; 'support' can easily lead to 'control' especially when the state gets involved in planning for women, and public power impinges on the private realm. It is hardly surprising, in view of subsequent events in the twentieth-century, that many women look with extreme suspicion on any form of collectivization of childcare, or intervention of the state in reproductive matters, education or health. This is because of past associations with political indoctrination, selective breeding, and eugenics (pp.59–61). Likewise, the co-operative approach to housekeeping was never generally popular with working-class women. In Britain provision of state nurseries has often been aimed at the working-class and 'inadequate'. State intervention and advice on childrearing has frequently given women the impression that they are being criticized (MacKenzie, 1989). It seems to me from observations and conver-sations that many working-class women are suspicious of any initiative in which middle-class women are in authority over them, telling them what to do, and 'judging' them. See-sawing: I view women who have authority over other women, especially those in the patriarchal health and welfare services, with great caution. The question of whether such elite are consciously exerting power over other women in their own right 'to keep other women down' and to retain their own privileged position, or acting as 'soft cops' as the agents of patriarchy, has barely been touched on in feminist literature (cf. Delamont, 1989).

Urban opportunities

Elizabeth Wilson (1991) draws out the contradiction, for women, of the city being seen as a place of freedom and opportunity economically and socially, but as a potentially dangerous place sexually. It was not a place to stroll as a bohemian *flaveuse* (Heron, 1993: 6). Freedom may really mean sexual freedom (and 'permission') for men to look for 'opportunities' (available women), which results in many women seeing the city as a place of potential fear and exploitation. Nevertheless the city, especially compared with the restrictions of rural life, offered women in the nineteenth century the chance to mix in a wider social and cultural circle. As I have already said, women of society acted as leaders of fashion and taste, having a major input in determining the location of fashionable residential areas. Women also had a greater range of employment opportunities in towns, and the likelihood of greater freedom of movement, compared with living in the countryside, especially with improved urban lighting and paving. Women could claim a legitimate right to be out shopping, which was a respectable activity associated with their homemaking role, thus having their own public space, albeit ordered around male retail business interests. The emphasis on building as a reflection of civic pride, and the concept of 'citizenship' (Beckett and Cherry, 1987; Dixon and Muthesius, 1978; Bookchin, 1992: chapter 8) was to women's advantage in creating public libraries, museums and art galleries, which women could use, without being accused of impropriety, for educational and cultural reasons.

Further zone zapping was needed to make existing cities more user-friendly for women, for, although shopping gave women a legitimate reason to be out in public, unaccompanied women might still be accosted even in such prestigious retail thoroughfares as Regent Street in London (Walkowitz, 1992: 129). But there was an underprovision of facilities and amenities specifically for women. In this section, I am stressing the shopping and mobility aspects to compensate for the emphasis on housing management, which is often presented as the only possible interest women might have had in town planning matters in the nineteenth century. The emphasis on industry, production and work has tended to detract from the question of consumption and retail development in the mainstream account of nineteenth-century urbanization. Of course, substantial numbers of working-class women worked in factories and mills. It was easy for some men to belittle such concerns when voiced by bourgeois women. But, the prosperity of the British Empire, and the continuation of the industrial revolution was dependent on trade, and consumers, that is shoppers (mainly women), were the ones who bought the manufactured goods and imported food supplies which kept the ships and factories in business. Every effort was made to woo the middle-class woman shopper (Adburgham, 1989; MacKeith, 1986). Working-class women knew they were not welcome in the new emporia, nor in the civic centre public buildings except as cleaners. It is significant that MacKeith's work (which earlier appeared in *The Planner*, Vol. 71: 9, January 1985) was the subject of a derogatory review (Curl, 1986), possibly because she was writing about matters on the wrong side of the public/private dichotomy.

Women's mobility was increased by the excellent public transport systems that were provided before the motorcar age, but it was restricted by the lack of public lavatories (WDS, 1990 'At Women's Convenience'). It was inconceivable for an unchaperoned, middle-class, woman to enter a restaurant or public house. Adburgham (1989: 231) comments that the provision of lavatories and tea shops (including Lyons) was the result of pressure from the women's suffrage group. The 1848 Public Health Act provided the first general powers for the provision of public conveniences. However, the emphasis was primarily on male provisions in spite of the activities of the Ladies Association for the Diffusion of Sanitary Knowledge. Whilst 'gas and water socialism' by men was praised, what was called 'sewers and drains feminism' (Greed, 1987) was seen as a joke in the nineteenth-century. However, despairing of waiting for state intervention, in 1884 the Ladies Lavatory Company opened its own private public conveniences at Oxford Circus, for ladies who had to spend the whole day in town (Adburgham, 1989: 231), and several department stores also provided facilities. Although public provision was poor, in those days women (not ladies) could probably breastfeed babies and children could 'go' in public without the looks that such natural activities cause today (Leach, 1979). Nowadays society has become much more fussy about natural functions in public, without providing adequate alternative solutions. By 1928 in London, there were 233 public conveniences for men, and 84 for women, containing 1,260 cubicles plus 2,610 urinals for men, and only 876 cubicles for women and the situation has got worse not better (and as well men have pubs and clubs, as discussed on p.43 and in WDS, 1990). It would seem that the municipalization of town planning, with its gendered agenda, detracted from the progress that had already been made through voluntary action and entrepreneurial initiatives to make cities better for women.

RESTRICTIONS OR REVELATIONS

Sex and space

Whilst women might welcome the greater freedom urbanization gave them, among malestream planners there seemed to be considerable difference of opinion as to women's position in urban society. In some of the more radical model communities (especially in North America, Hayden, 1976), women might be treated at one level as equal co-workers and believers, but at another level might be seen as property to be collectivized and shared according to ideological principles. Even if they were seen as individuals they might still be expected to demonstrate their liberation and lack of possessiveness or selfishness by rejecting monogamous relationships, in the name of free love or antinomianism. Kumar (1991: 30) refers to de Sade's *La philosophie de la boudoir* of 1795, with its castigation of monogamy and individualism, as being reflective of certain mainstream utopianist attitudes. Goodwin (1978) notes that Fourier substituted psychology for religion. He prefigured belief control by the state in setting up what he called '*la création sociale*', based upon the phalanx community (Goret, 1974). Religious communities such as the Mormons emphasized polygamy, whilst the Oneida community had some peculiar practices based on 'sharing', although they also produced world-famous cutlery and had women-centred kitchen houses as the nucleus of each community. Sheila Jeffreys (1990: 130), in her exploration of 1960s sexual liberation, shows the influence of these early town planning experiments in sexual freedom, on such well known figures as Dr Comfort, of *Joy of Sex* fame (1979). He experimented on similar lines in his Sandstone community in California, indicating a direct line back to the practices in early Oneida communities. Likewise, H. G. Wells, influenced by Plato, visualized his 'samuri' in his *Modern Utopia* of 1933 as free lovers in one 'group marriage', drawing on the example of the Oneida community (Brandon, 1991: 171).

In contrast, in mainstream British town planning, importance was generally given to domestic ideology which emphasized the formation of individual, exclusive households, with the man as breadwinner and wife as homemaker. This model was to inform the nature of urban policy-making in respect of land-use zoning, which separated employment from residential areas, when town planning became a municipal function in the early twentieth-century (see chapter 7). But neither of these extremes, collectivization into the public realm, nor isolation into the private realm, were ideal for women. Both might be seen as being manifestations of the eroticization of power differences, based upon the subordination (and humiliation) of women, and the conquest of space. Jeffries (1990: 308) stresses the sexual importance of establishing difference, division and domination in public policies that impinge on private lives (compare Barker-Benfield, 1972). Such power restrained and policed men as well as women, gender differences being set in concrete into urban spatial structure to prevent the destabilizing of public/private, male/female dichotomies (Sandercock and Forsyth, 1992: 53). There was no space for intermediate, shared zones not proscribed by gender.

Spirit and space

Some women sought to transcend such power relationships and related urban dichotomies. Hayden, in her research on seven American utopias (1976: 33) in which she studied Shakers, Mormons, Fourierists, Perfectionists, Inspirationists, Union Colonists and Llano Colonists, scans the scene from a belief perspective, rather than a purely feminist one as in the case of her later book (1981). Many of these communities were organized spatially by town design, and aspatially through social and managerial structures around the religious principles which they espoused. Rosabeth Moss Kanter, who many years later became the female guru of the corporate business world, studied for her doctorate the ways in which power was maintained in nineteenth-century utopian and 1960s hippie communities (Kanter, 1977). She found that the projection of convincing, belief systems was the secret of power over others, enabling the creation of total realities, which could then readily be embodied in bricks and mortar (1972). Arguably, she used her findings in the service of promoting the illusion of the enterprise culture in the 1980s, possessing none of the English reservation about the non-niceness of trade. It seems to me that she believed sincerely that commercial, rather than socialist, cultural change would promote feminist ends, thus seeking to transform and use capitalism, by creating more caring business cultures (effectively zone-zapping). David Walker significantly comments, 'most of all she is a prophet' (*The Times Higher Education Supplement*, August 1992: 15, No.1034).

The communal kitchen rather than the men's workplace was at the centre of such communities. Anne Lee, founder of the Shakers in 1776 (women leaders are a feature of many utopian communities), believed in the concept of heaven on earth and initiated 'Shaker spatial discipline' governing planning and architecture. Whilst condemned by critics as 'petty', it is no more so than modern development control under planning law, and a good deal more useful. Shaker furniture and architecture are valued nowadays, and the style may be seen as a precursor of the functionalist movement in which 'form follows function'. In contrast, the Shakers coined the proverb, 'every force evolves a form' (Hayden, 1976: 32). Their spatial environment was prescribed by their spiritual beliefs, or by 'the spirit' – the force of life. This is an example of the ultimate in reproduction over space of social relations, rather than spiritual relations in which the gnostic spirit/body dichotomy was transcended. The aim might be expressed as, 'life in the presence of God . . . in the ordinary world bearing the holiness once associated with sacred space and time' (Green, 1986: xii–xiii). The community itself, not some spatial object, or place, was seen as sacred, thus breaking the sacred/profane dichotomy at source, and making redundant (some of) the zoning which derived from it (Hayden, 1976: 32). Many visionaries was recourse to divine power as the only way to break down patriarchal dichotomies and find a way out (see the classic zone-zapping text, Galatians 3: 28).

7

PROFESSIONAL POWER OVER PRIVATE SPACE

THE PUBLIC ORGANIZATION OF PLANNING

Public and private realms

This chapter looks at the situation in the first half of the twentieth century, when modern town planning became a profession and a function of local government. In the process, planning and planners were reified as male. First, I will consider women's position within the public realm of planning practice, and the limited opportunities this gave them to act as zone zappers. I will then discuss the beliefs which informed planning policy and the effect this had on city form. Chapters 7 and 8 are broadly chronological, but greater attention is given to the planning of the private realm in this chapter, whilst in the next the public issues, such as employment, are emphasized.

The creation of a profession

In spite of its previous associations with utopianism, feminism and other alternative social movements, town planning took on a sanitized, no-nonsense persona in the early twentieth century. It became a respectable malestream profession rather than a visionary social movement. The Town Planning Institute held its inaugural meeting in November 1913 (Ashworth, 1968: 193). By May 1914 there were 52 members, 18 associate members, 6 legal members, 28 honorary members and 11 associates (as described by Cherry in *The Planner*, Vol. 70, No.9: 8–10, September 1984 and see my footnote 4, p.17). Incidentally, Cherry's article is one of a set published in the 70th Anniversary edition of *The Planner*, headed by a cameo photograph. This shows twelve Edwardian men standing in a field, poring over a site plan, with one woman (only) standing on the edge of the group trying to peer at it too. 'Professionalism was in the air' (Thompson: 1968; Greed, 1991: 56); it was fashionable to set up committees and thus create professions. The membership was overwhelmingly male, but there was one woman on the committee – Henrietta Barnett. The TCPA (Town and Country Planning

Association) was established in 1903 as successor to the Garden Cities Association and constituted the main 'alternative' planning pressure group. It had three women out of twenty-three members on its founding committee – and one of these was a Mrs. E. Howard (Hardy, 1991a: 83). But, Bristol was the only local branch (*ibid.*: 164) to be chaired by a woman, Miss Hilda Cashmore, a member of the Bristol Women's Advisory Housing Committee. In spite of the participation of women in the early town planning movement, Hardy (1991a: 84) describes women's role in the early TCPA as 'ambiguous' and suggests they were concerned primarily with health benefits to children, and the establishment of 'girls clubs'. A separate Woman's League had been established in 1903. Similarily the RICS set up a separate 'housing manager' section for women (Greed, 1991: 62) when the 1919 Sex Disqualification (Removal) Act allowed women into surveying, to keep them off the main pitch.

In the early years there were very few women in the Institute. Those there were, were either associates – that is those in the non-professional membership category – or full members who were likely to be architects. In the beginning most members were prior professionals, that is they were already qualified as architects, surveyors or engineers. Architecture was the only built environment profession which already had a small female membership. The idea of a woman becoming an architect, as against a surveyor or engineer, was relatively more acceptable because she could be safely categorized as a 'women in art and design' (Anscombe, 1984; Attfield and Kirkham, 1989). In 1898, Ethel Charles became the first qualified woman architect, although there were several 'unqualified', competent ones around, according to Lynne Walker the architectural historian and others (Wigfall, 1980). In America there seemed to be more opportunities for women to become architects, such as Julia Morgan who undertook large-scale commercial projects. By the 1920s in Britain there were several women's architectural partnerships such as Norah Aiton and Betty Scott, famous for 'modern' office designs and factory buildings, for example in Derby. Other women entered surveying; Irene Barclay was the first to do so (Barclay, 1980, Greed, 1991: 61).

As I said in chapter 1, it is essential to 'name' women planners to redress the imbalance in the malestream historical record. The names of Mary Barson, Elizabeth Denby, Elizabeth Halton, Elizabeth Ursula Chesterton, Nora Dumphy, Mrs Abrams, Betty Yule, Annette Cole, Caroline Gilson (ARIBA), and Eileen Benfield (a planning assistant in Glamorgan) feature in the early planning journals. Most of these women were only mentioned when they first joined in lists of new members. However, Miss Adburgham was frequently mentioned in her capacity as a member of the education and membership committees in 1934, and later became a member of the general (ruling) committee. Scandinavian women sometimes feature for example as Rosemary Stjernstedt in 1946, (Vol. 32, No.6: 225) who wrote an article on the Greater

Stockholm plan (unless I indicate otherwise, all journal references are to *The Planner*). Some women crop up as authors or reviewers of books, such as Margaret Blomfield who wrote *Our Towns: A close up* (1943). An obituary of Betty Benson, AM, BA, ARIBA, who was an assistant architect at Coventry is given by Jane Drew in 1946 (Vol. 32, No.6: 204, July/August). Women seemed to have a cosy and private rather than confrontational public place in the Institute. For example, there is a letter from Leila Jenkins in February 1944, ending 'trust you are all well' (Vol. 30, No.2: 128).

Potential zone zappers in the built environment professions in the early days included: mavericks, such as avante-garde women architects involved in the art world; marginal women, who had not been conditioned to accept a conventional gender role; committee women who might be seen as an early form of femocrat; women from reformist and religious backgrounds; and entrepreneurial business women (cf. traders, pp.81–2). In planning there is less opportunity for private practice than in surveying or architecture. However, in the aligned area of producing town maps Phyllis Pearsall established the Geographia A-Z maps company, in the 1920s personally mapping much of London (Pearsall, 1990). I found little connection between the early twentieth-century world of town planning and first-wave feminism. The two realms seem to have been quite separate, except for the appearance of a few of the same individuals in each, such as Henrietta Barnett. Occasionally women from North America materialized in the journal, for example Theodora Kimball of Harvard who was made an associate of the Institute in June 1923 (Vol. 9: 162). She later married a fellow planner with whom she produced a major planning book (Kimball-Hubbard 1929). Margaret Cole is featured in June Purvis's book on women in education (1991: 90) in relation to her experience at Roedean (1907–10) and appears as unacknowledged co-author of a book in a series on economic and town planning with her husband, who admitted (Cole, 1945: in the frontispiece) 'a number of these books have been written in collaboration with my wife'.

Municipalization of planning

Under the 1909 and 1919 Housing and Town Planning Acts local authorities were required to prepare town schemes (plans). But their powers of enforcement were limited, and there was considerable opposition from landed interests. The sort of people who ran the new town planning system were unlikely to be visionaries, as they were usually men drawn from the pre-existing built environment professions and institutional frameworks and thus were, arguably, imbued with traditional patriarchal values. The largest single profession represented at the launch of the TPI in 1914 was architecture, with engineering and surveying not far behind. As late as 1971 prior professionals constituted 45 per cent of RTPI members, and were more likely to have professional fathers than graduate planners according to Susanna Marcus

(1971) in her famous article 'Planners who are you?' (Vol. 57: 54). The planning assistants who operated the system (p.30) were unlikely to be professionally qualified and came from a municipal adminstrative background, or from technician grades.

The examinations system in the built environment professions was still evolving. In comparison with planning, some surveyors thought the expertise required was too practical to be capable of being examined (Greed, 1991: 57). Few of the middle-class held formal, professional or academic qualifications, employment depending more on 'who' rather than 'what' a man knew. Sir Frederick Osborn (a great planner, even a philosopher king) is an example of this. He was born in 1885, worked as a city office boy at 15, and through his subsequent experience in housing management became involved in the garden city movement, unlike many women who found housing management a dead end. He met Ebenezer Howard and other useful people along the way (Vol. 42: 209, July/August 1955). The TPI was a learned rather than a qualifying association at the beginning (Millerson, 1964: 83). There was pressure for formalization of qualifications from men in local government seeking professional status. The emergence of the planners may be seen as one manifestation of a general embourgeoisement and professionalization of the growing male, white collar workforce, during the period of social and economic change after the First World War (Marwick, 1986). Those planners who were prior professionals or from elite backgrounds saw the new planning system as a means to exert power. Chameleon-like, they took on the technical persona of local government officers, when some might have been more realistically described as philosopher kings or high-level managers (pp.30–2). Of Thomas Adams, the first President (Cherry, 1981: 41), it was commented in *The Planner* (Vol. 36, No.1: 7, 1949–50) 'having sown his wild oats as a private practitioner he did not chose freedom but the yoke and joined local government service in 1935 as chief planning officer'. Likewise, Clough Williams Ellis, born in 1883, an arty, urban-design type planner and builder of Port Meirion, was later chairman of the First New Town Development Corporation at Stevenage in 1946 (Vol. 44: 76, February 1958), taking on the persona of a socialist scientific planner with the job. The attainment of formal town planning qualifications was likely to be a requirement for career progression only for lower status staff: the technician planners who ran the system (pp.30–2).

The creation of a cadre of public-sector planners appeared to detract from the efforts and priorities of those women planners who were active in the private realm of voluntary housing and town planning work. Their expertise was not formally recognized, for few had any official qualifications. They were not in full-time paid employment, and so they were likely to be invisible. There seemed to be an intense dislike of 'lady bountifuls', who were seen as taking away middle-class men's jobs by doing professional work for free. One never hears of lord bountifuls. Aligned to town planning, housing management had been the one area of the built environment professions where women were actually in the majority, but with increasing municipalization women were progressively pushed out. It seems to me, as I describe in Greed, 1991: 62–4 in respect of housing management, that bright, maverick women often start a social movement or new entrepreneurial area of professional activity and then dull men gradually take it over, excluding the women. Interestingly, though, a few patriarchal non-visionary women are usually retained, possibly to block the path of more unsettling women.

Philosopher kings, or reference to their works, also had an important role to play. Foreign ones, ideally with unpronounceable names, and elderly or dead ones, gave a mystique and status to the new profession. Their visionary ideas and genius were used to legitimate the importance of the state planning system. It is doubtful whether some of the planners in local government had ever read their works. An element of selection and filtering occurred, as the new leaders of planning took only those ideas from the nineteenth-century town planning movement which legitimated their approach to planning and left out the more controversial or politically unsettling aspects. The land-use side of Howard's theories were retained whilst his more visionary social and economic ideas were down played. Howard had died in 1928. The work of some geniuses was ignored altogether: women visionaries (as described in chapter 6) were not venerated as philosopher queens and were not even seen as important. Having been rejected as unimportant at the beginning of the century (and of the profession) they were unlikely to regain status. An RTPI teaching unit (RTPI, 1986, Unit 2: 25) is at great pains to stress that, 'despite her undoubted stature, Octavia Hill is not a major figure in the history of town planning'. In contrast, on the next page Robert Owen *is*. He is described as a major figure in town planning precisely because of the diversity of his interests: the very characteristic which goes against Octavia. Although Hill is seen as 'only a housing manager', her ideas were influential across the fields of urban, rural, and regional planning policy (Darley, 1990).

Pre-war planning education

Before the Second World War one could only study town planning as a postgraduate subject, and courses, few though they were, were mainly part-time. It was expected that candidates would combine their studies with professional employment, a potential exclusionary mechanism for women, of whom, as far as I could ascertain, there were very few enrolled on such courses. The first chair in town planning had been established at Liverpool University in 1909, by William Lever of Port Sunlight fame, under the professorship of Stanley Adshead. He set up the famous MCD course (Masters in Civic Design) (Ashworth, 1968: 193; Adshead, 1923), which was architecturally based (Vol. 25: 14 November 1938). Topics included civic engineering, hygiene, civic law, civic architecture, and landscape design. An article by Professor Allen (Vol. 34, No.2: 31–42, January–February 1948) proudly states that the subjects barely changed over 30 years, they were so good. But some apparently odd subjects crop up in accounts of early courses such as 'public food distribution': a faint reflection of feminist co-operative housekeeping perhaps. Although in a sense everyone on the early courses was elite, Allen comments that he saw a distinction between what he called the specialist planner or administrator (my technician planner) and the planner who had a stronger humanities/social sciences bias (my philosopher king). Articles from the journal include photographs of pipe-smoking young gentlemen in sports jackets gathered in huddles around drawing boards. The emphasis upon studio-based, group projects has continued through planning education to the present day. Seesawing: the male, adverserial 'crit session' approach to project assessment, an import from architectural courses, continues to cause problems for women. Also, the tradition derived from geography of going out into the field to do a scientific survey survives. The 'field' might be the local residential area, or shopping centre, but for some men this may be like going into the unknown, crossing public/private realm barriers into the world of women or working-class people.

UCL (University College London) was one of the most prominent pre-war colleges (Vol. 25: 128, February 1939; Collins, 1989). Other courses were established at Birmingham (under Cadbury's patronage, Gardiner, 1923); Durham (later moved to Newcastle); Manchester; Edinburgh; Nottingham; Leeds; and Regent Street Polytechnic, London (a first-wave 1930s Poly, Venables, 1955). There were strong links between practitioners and colleges, for example at Leeds, where Desmond Heap the famous planning lawyer (cf. Heap, 1991) was the law lecturer as well as being employed by the town council. Many of these historic colleges contain the elite planning courses today and have produced a significant number of notable women planners. See-sawing: a course leader at one such college wrote to me to the effect that there was no reason to suppose there were any significant gender variations in the career destination of his M.Phil students, because several of his former female students were in charge of London Boroughs, suggesting that gender does not hinder advancement. Class and brains help where gender fails!

Women's place in planning?

In theory women were equally as able as men to join the TPI and to work as planners in local government. The 1919 Sex Disqualification (Removal) Act enabled women to enter the professions. In practice, for women there was limited access to qualifying educational courses, problems of finding professional articles, and discretionary recruitment and employment policies. On marriage a woman was expected to give up her job, whether she had children or not. Therefore the idea of maternity leave did not even come onto the (men's) agenda as it was strictly a private-realm issue. But the principle that men should be given their jobs back after military service and that their absence should not go against them in terms of promotion, pensions, and salary rights was well established by the time of the First World War because war was a public realm issue. Women may not have even been considered as a valid pool from which to draw employees, not being seen to possess the right credentials according to mainstream criteria.

One often gets the impression that the first wave of urban feminism died away in the 1920s along with the suffragette movement after women had achieved the vote (Banks, 1981: 118). Rather, it would seem that certain types of women were still actively involved in the town planning movement, but in voluntary groups, not as professional planners. Acceptable women even sat on governmental committees dealing with housing and town planning legislation (Ravetz, 1989: 187–205). On these committees women's expected role was not to act as a zone zapper by proposing town planning schemes which would break down divisions between work and home. Rather, 'women's views were seen as important but only in their capacity as housewives' an observation Ravetz (1989: 199) cites from MATRIX (the women's architects group) (1984: 74). Thus, women planners, although occasionally invited to participate in the public realm, were to restrict themselves to discussions of the private domestic realm. Ravetz mentions the work of Margaret Bondfield and of Eleanor Rathbone in this context. One needs to look at bodies such as the Electricity Women's Group, the Electrical Association for Women (and the related Electricity Distribution Association), and the Women's Gas Council (and later Federation, Beauchamp and Hudson, 1985) to find women and planning still going strong in the missing years. They were dealing with questions of infrastructural services provision which was a major determinant in the future shape of both town planning and house design (Ravetz, 1993). Women were still concerned with sewers and drains, an old feminist issue (Greed, 1987). Lady Pepler expressed concern about the nation's drains, with support from the National Federation of Women's Institutes, which published a book by Cicely McCall, *Our Villages* (1958), stating that there were 3,850 villages without drainage. Joan Monypenny of the British Commercial Gas Association gave a talk to the TPI on 'planning and the gas industry' (Vol. 30, No.5: 27, November–December 1943).

Women had not been allowed to vote or stand in national elections until after the First World War, but they could vote in some local elections, and stand for election for public office provided they were property owners in their own right (cf. *The Girls' Own Paper*, 1904). For example, Charlotte Despard was the first woman poor law guardian in 1906 (Purvis, 1991: 122), and both Beatrice Webb and Henrietta Barnett were involved in local government committee work. Other women were poor law guardians (albeit only 136 out of 28,000 in 1892) or sat on school boards and parish councils (Levine, 1987: 72). The first women involved in the administration of the town planning system were mainly upper-class elite people, who acted to reinforce the class structure (Stacey and Price, 1981: 67). They had the right cultural capital (Delamont, 1989), but limited knowledge of working-class women's lives. Even Octavia Hill considered that one cold-water tap per floor was perfectly adequate for her working-class tenants. The tradition of using upper-class women in senior posts persisted, even into the post-war period, when town planning was meant to be for the working classes and was run mainly by middle-class men who espoused socialist principles. Perhaps upper-class women were seen as a tried and tested type of acceptable woman who would not break the patriarchal mould. In May 1950, Lady Evelyn Denington, became vice chairman of LCC housing committee and chairman of the planning committee at St Pancras, and subsequently was one of the few female members of a New Town Development Corporation, in Stevenage. Some titled ladies feature in central government planning roles. Baroness Sharp, is frequently mentioned, and was at the opening of SAUS (School of Advanced Urban Studies) in Bristol in 1973 (Vol. 59: 193), Baroness Serota became chairman of the committee for local administration in England and Wales in 1977, and Lady Young gained recognition as an authority on town planning. See-sawing to the present: I have observed several examples of senior men planners of working-class origin in local government promoting upper middle-class women, creating power alliances of left-wing men and right-wing women which would seem to block the progress of more radical or feminist women (Greed, 1993b: 268).

In spite of the involvement of a few women in local government committees, some urban feminists see the importance of this factor as being far outweighed by the development of a paternalistic style of municipal management in local government. This was based on traditional rather than radical socialism, which was seen as highly patriarchal, resulting in gendered planning and housing policies. London County Council became dominated by a paternalistic Labourist model of local government, in which an uneasy alliance existed between Labour leaders, professional men and upper-class women. A new middle-class of salaried municipal clerks and technicians came into being who made a living out of ostensibly solving the problems of the working class. Barnes (1989) sees this top-down, Morrisonian model of local government as being a major hindrance for women. Herbert Morrison was secretary of the London Labour Party (1915–47), chairman of Lambeth Council, and of the LCC (1922–45). This strikes a chord for me as a Londoner. But some colleagues tell me that I judge the LCC too harshly, and that it was a progressive body responsible for the construction and management of a wide range of schools, state housing schemes and public works, with women such as Evelyn Denington being vice-chair of the housing committee and chair of two of its sub-committees (Roberts, 1991: 107).

TOWN PLANNING POLICY

Belief?

At first glance, when I looked at the nature of planning policy in the first part of the twentieth century, it seemed, to me that, unlike the Shakers (p.106), who believed that 'every force evolves a form', the early twentieth-century planners had lost the inspirational force and were left only with the form. An over-preoccupation with quantitative density standards appeared to be all that they had left – an empty shell – once the utopian and visionary beliefs which had previously underpinned planning were abandoned. But I realized these ostensibly scientific standards were manifestations of 'forceful' patriarchal beliefs as to how the world should be. Belief in public/private dichotomies could be enforced by land-use zoning policies. Emphasis was given to the principles of decentralization and dispersal, which were linked to the density debate, in order to achieve so-called balance. Women's perceived role was to be expressed in the design of residential neighbourhoods, especially in public (council) housing. But the only direct reference to belief that I could find in the early journals was that members of the Institute attended a service at Canterbury Cathedral, during one of their first conferences in October 1925 (Vol. 11, No.12: 311).

Dispersal and density

One of the arguments for planning at the turn of the last century was that cities were overcrowded. This legitimated the need for decentralization of overspill population and the creation of counter-magnet satellite garden cities. Morrison favoured urban decentralization (Ward, 1991) with an emphasis on moving out the working classes. Barnes (1989) comments that this was problematic for women in particular, creating inconvenience and reducing local employment opportunities. The clearance of slum areas, which usually accompanied such initiatives, destroyed the so-called clutter which made inner urban life tolerable and convenient. Like a flower bed, the city had to be 'thinned out' and 'weeded' by slum clearance and population relocation. Consider parallels with the controversial 'weed and seed' policy adopted in Los Angeles following the riots in 1992. British policy was highly eugenic and judgmental in character. To undertake 'spatial eugenics' the planners had to indulge in complex calculations to come up with the 'correct' density for the healthy development of an area to prove overcrowding (over-population), thus legitimating their power to move people out (Gibbon, Vol. 12, No.2: 25, December 1925). The links between density, eugenics, racism and function-alism can be found in a range of urban theorists' work as diverse as Camillo Sitte in *Der Städtebau* in 1889, to proponents of the Chicago School of Sociology, especially Homer Hoyt (cf. Bulmer, 1984: chapter 5 on 'the negro'). Sitte's heroes were Darwin, Beethoven and Richard Wagner, for whom he designed the stage sets for the first performance of *Parsifal*. Sitte saw the evolutionary struggle for survival as an essentially 'germanic concept' (Collins and Crosemann-Collins, 1965: 15). Geddes was influenced by eugenics, and by Darwin's ideas of 'selection', and promoted 'the idea of civics and eugenics in association' (Geddes, 1915: 388). Matless (1992: 466) demonstrates Geddes' use of Freudianism to support eugenic town planning. One can see 'male' sexual allusions in the metaphors used to describe urban problems in discussions of retention and uncontrolled sprawl, and denseness being associated with tenseness. Likewise, images of urban form were strongly sexualized, as reflected in the planners' pre-occupation with controlling linear ribbon development, and, of course, in the later fixation with high-rise building forms (cf. Barker-Benfield, 1972). Thomas Sharp's work also manifested negative constructions of town/country; black/white; male/female dichotomies. Seesawing: Matless notes, that Christopher Fagg, the geographer and follower of Geddes, had a long correspondence with Freud, and pioneered urban and regional survey methods based on eugenic and evolutionary theory, using in 1923 what was to be my home area of Croydon, and the county of Surrey.

Eugenic questions of population were never far from the density debate, with population forecasts swinging widely in predicting 'too many' or 'too few'. In an article in 1939 (Vol. 26, No.1: 6, November–December) 'population tendencies' were seen as stationary and about to decline, and the question of 'replanning for depopulation' was discussed. Roberts (1991) comments that planning was pro-natalist at this time. This was purely an abstract numerical sentiment, for as Roberts points out, residential areas were hardly planned in a manner convenient to the needs of women with children. Aldous Huxley (a philosopher king) in his article 'Health for All' in a special edition of *Picture Post*, 4 January 1941, entitled 'A Plan for Britain', also expressed pro-natalist views, as did the architects of the Welfare State (Lake, 1941). But pro-natalist policies were selective as to who was 'fit to breed', with considerable condemnation being reserved for working-class parents for 'causing' the post-war baby boom. There were many articles on the 'density debate' in the early planning journals such as one in March, 1927 (Vol. 13: 149) which discussed how to 'remodel and reconstruct the congested interior of the town'. More famous is the seminal article by Barry Parker, 'Economy in Estate Development' in July 1928 (Vol. 14, No.8: 177) which includes the diagram captioned, 'nothing gained by overcrowding' which demonstrates that lower density development might take no more land than terrace development. This article on density is significant in combining traditional 'aesthetic' design aspects, and the 'scientific' mathematical aspects of planning, which appear virtually identical as to reasoning and purpose. The 'density debate' provides the cusp, the spinode, the cross-over point, between the old and new approach to town planning. Density also became the basis for control as a legal tool in the enforcement of town planning policy, because it is quantifiable and thus incontestable as a means of justifying separation. Formal planning law took over the control role, which intuitive 'design' had previously exercised. An endless sequence of planning law books began (cf. Willis, 1910) which established the power of the lawyer over this new-fangled form of land law, as another component of real property law, an ancient form of patriarchal control (Holcombe, 1983).

Biological zoning

As indicated, early twentieth-century town planning theory was influenced by evolutionary theory, Freudianism and biological principles being applied to space. Patrick Geddes, biologist turned town planner, saw urban life as being encapsulated into a tripartate division; namely into 'place, folk, work', (Boardman, 1978). Susan Kingsley Kent (1987: 35) reports that in respect of opposing women's suffrage, Patrick Geddes and J. Arthur Thomson (a fellow biologist) commented in *The Evolution of Sex* (1889) that, 'what was decided by the prehistoric protozoa cannot be annulled by act of parliament'. Other models included Le Play's, 'man, place, work' (Doxiadis, 1968: 22), but also body/folk/home on the 'female' side of the dichotomy. Such divisions hark back indirectly to ancient, gendered, philosophical concepts of 'mind, body, soul' (Descartes, 1637), derived from Artistotle's 'knowing, doing, feeling' (Patterson, 1992). Such divisions never fit woman, who is seen as an untidy, semi-human mixture of procreative nature and emotional 'spirit' which goes beyond the objectivity of male *intuis* (Macleod, 1991). But how do such trichotomies fit in with the public/private, male/female dichotomies stressed in this research? It seems to me that one must make the male/female division first and take these trichotomies as subcategories which encapsulate how reality is 'seen' from a 'public' perspective in which the world is presumed to be populated chiefly by men. *The Planner* was full of phrases such as 'town planning and the man in the street' (Vol. 16, No. 6: 143, April 1930). In spite of the claim to a scientific approach, in looking at the textbooks of the time (few though they were) one is struck by how amateur, gentlemanly and full of home 'truths' they are (such as Adshead, 1923; Abercrombie, 1943). Other planning heroes, such as Mumford, Osborn and Pepler, enforce the image of a male centred world (Cherry, 1981: the cover entirely comprises pictures of male planners).

Natural zoning divisions did not evolve of their own accord. The planners had to intervene through corrective zoning (Vol. 25, No.7: 232, May 1939). Some planners were intent on projecting such beliefs onto the wayward built environment to make it conform to their image of reality, but they encountered considerable opposition imposing their ideas. Men planners are not a unitary group and some appear to be more sensitive and enlightened than others. For example, Professor Cordingley (Vol. 25 No.1: 11, November, 1938) asked if 'remorseless' segregation of industry and residence was really necessary. He stated that segregation implies that 'work is obnoxious'. The growing emphasis on 'functionalism' in the town planning movement could work to women's advantage. In April 1932 (Vol. 18, No.6: 153) an item commented 'homes should be near workshops', for efficiency's sake, as if occasionally real-life practical considerations surplanted the patriarchal taboos about keeping women out of men's work realms. McCallum (1946: 184), referring to a photograph of some Swedish flats, commented

> this shows an almost perfect relationship between home and work, both within walking distance but clearly separated and carefully sited. Although this relationship is not always possible or necessary, it is essential that there should always be clear separation, and speedy, cheap and comfortable transport.

But such men did not win the power struggle against the zoners.

Planning inside housing

Whereas in ancient Greece the private realm of the household (*oikos*, p.80) was a respected separate zone (Boulding, 1992: I: 227), in the twentieth century the private, domestic realm was increasingly interfered with by public bodies. Thus any separate autonomy which women might have gained was restricted (Kolmerton, 1990). The Housing and Town Planning Acts of 1909 and 1919 not only introduced a rudimentary form of town planning control, but gave the planners an unprecedented opportunity to meddle in the private, domestic realm by prescribing the design and internal layout of social housing for the working-classes. Willis states that the working-class household was legally defined in 1903 for the purposes of eligibility to state housing provision, as consisting of someone who earnt 30/- or less and those residing with *him* (1910: 9), with specific categories of male, manual, employment being identified as valid. The extensive house-building programme introduced under the 1919 Act aimed not only at providing housing for the working classes in general, but at providing 'Homes for Heroes' (Swenarton, 1981) for soldiers returning from the First World War. Such definitions gave a 'bias' to the legislation which leaked across into the planning side of the acts, marginalizing women. The worst combination of conflicting perspectives occurs when upper-middle-class men designers operating from the public side of the public/private dichotomy impose inappropriate ideas on working-class women's domestic lives. Such designers might have been well-intentioned, but they might hold unrealistic, romantic images of the artisan family, and have little idea of the realities of daily life. They might see housing estates as townscape or layouts from an urban design perspective in which the designer is above, and outside, and not himself part of what he is designing (p.69).

Raymond Unwin (Creese, 1967: 17) felt perfectly justified in interfering in the internal design of the artisan's cottage, breaking down the internal divisions between the living room and parlour. He thought such separate rooms for the working class family pretentious, and favoured one large room, suggesting that individuals might gain privacy by sitting in the inglenook! He ignored the fact that many families needed a public front zone inside the private dwelling – the parlour where they could entertain guests and strangers, without having to bring them into the private zone of the back room of intimate family life. This is an example of public/private divisions appearing like interlinked nested hierarchies overlapping in each other's space (rather like Russian dolls). He suggested the stove might be put in the living room with 'perhaps a small washing up sink for the crockery' (Creese, 1967: 48). This arrangement reflected his admiration for the manorial hall, which held little furniture except for a large table in the centre. Feminists have criticized this arrangement (Roberts, 1991: chapter 2). It prefigured open-plan living of the 1960s (MATRIX, 1984), which did not give women a space of their own to escape, nor did it allow for so-called domestic clutter.

Woman was primarily seen as a housewife and mother, who presumably did not work outside the home, but who served the needs of her husband the breadwinner. There was no intellectual space in this model for the reconceptualization of women's and men's roles within the domestic realm, nor for the redesigning of housing provision on co-operative houskeeping lines. The emphasis was on building housing estates, not on creating balanced model towns with the full range of land uses and facilities. But, adaptions of the garden city style of artisan cottages were used for the new housing estates. The designs of the architects Unwin and Parker were readily copied to form the basis of the Tudor Walters (1918) housing standards for the new council housing. The end result may be seen as constituting an outward stage set. The spirit and function of the utopian and communitarian ideas behind the garden city had been abandoned. Urban feminist ideas of co-operative housekeeping could be twisted for use against women, to give the impression that it was selfish to be a housewife, presumably doing private family tasks rather than public work. An article in February 1933 (Vol. 19, No.4: 72) comments that 'individual housekeeping is an inadmissable luxury', but offers no alternative. The theme of co-operation, even provision of nurseries and laundries, as subsequently appeared briefly in some LCC schemes in London, might be seen as a way of increasing women's participation in production as a reserve labour force, but not in women's control over their lives and families. Women were clearly inefficient, as evidenced by an article which applied Taylorist time and motion principles to women's chores (Taylor, 1895). For example an item (Vol. 13: 117, February 1927) illustrated by diagrams of foot movements, discusses the amount of time it takes to prepare tea, in a kitchen devoid of inquisitive children and hungry cats.

Nowadays (as raised in chapter 2) planning is more concerned with external layouts and city form, and thus its domain stops at the front door. Interference in the private zone in the early twentieth century was justified by appealing to planning's physical purpose, whilst today the same reason is used to exclude the interior zone. Also it was, undoubtedly, partly because the planners were dealing chiefly with social housing for the working classes – a captive population – that they could extend their power to the internal zone. In contrast, nowadays, with the increasing emphasis on the need for a mortgagariat, rather than a proletariat, to keep the economy going, private domestic consumption rather than public investment is encouraged. Nowadays the private realm of the insides of houses is generally out of bounds to planners. But, the public/private dichotomy which divides public council housing from private owner-occupied property is still crucial in determining the planners' attitude to different types of residential development. However, nowadays there is a dichotomy-defying exception to this rule. If a privately owned historic residential property is considered to be of exceptional historical or architectural public importance, it can be subject to urban conservation controls, inside as well as out (Greed, 1993a: 163).

Private housebuilding

Some could always afford not to be subject to public interference in the private realm of house and home. A tremendous amount of private house building was occurring, decentralization being encouraged by the growth of the motorcar. In 1910 90 per cent of all housing was rented, with owner occupation being limited to the more affluent classes (Swenarton, 1981), but between 1930 and 1940 alone 2.7 million houses were built, many of which were owner occupied (cf. Legrand, 1988 for the sweeping nature of twentieth century change). Planners sought to control unauthorized private sprawl through a series of inter-war planning acts, but many developers ignored the legislation as the penalties were minimal. The population could no longer be conveniently dichotomized into two: 'us' and 'them': the upper class and the 'planned' working class. A whole new 'uppitty' *middle* class was arising, which had taken town planning into their own hands, and had voted with their feet and set up home in the suburbs with the help of private-sector housebuilders. However, paradoxically, the middle-class should not be assumed to be zone zappers, elements of it being characterized by an obsession with social conformity, respectability, and making a distinction between itself and the working class. Private property developers, arguably a modern variation of the trader caste, had far greater power than the planners over urban development within what was still a *laissez-faire* situation, for all the planners' illusions that it was not. Interestingly, whilst capitalism is often seen as bad for women (and men), private enterprise house building, because it has to respond to customers' demands, might be seen as being potentially more reflective of women's needs than public sector socialist housing provision (or more exploitative?).

It was fashionable to belittle suburbia (p.2, and Priestley, 1934), but there seemed to be some ambiguity as to what form politically correct housing should take. McCallum (1946: 184) saw the detached house as an 'extravagance', and by means of a series of little diagrams suggested that families with children over the age of 1½ years old only should be allowed to move into larger houses. Amusingly, the bungalow, which in Britain is often seen as the epitome of petty, private-realm domesticity and lower-middle-class values, was seen in California as an extremely masculine *non-domestic*, public-realm form of development associated with roughing it, and as a relative of the log cabin and the ranch house. The bungalow was described as 'a socialist response to capitalism' (King, 1984: 142, and caption to plate 68). Regardless of what the philosopher kings and experts thought, many women found speculative, suburban housing to be designed in a way which reflected their requirements more closely than so-called social, state housing, especially because a separate front parlour was usually included (Ravetz, 1989: 198). This provided a space where they could relax away from their workplace, the kitchen. Although the suburban house itself was liked, there were problems with the decentralized location (this paradox is raised in chapter 2). There was a need for local amenities; indeed some private estates included shopping parades, developers having no qualms about the importance of 'consumption', unlike their public-sector counterparts.

OUTSIDE INFLUENCES: THE NEW TECHNOLOGY

It was not just British garden-city-type town planning which manifested a patriarchal bent in the early twentieth century. European, high rise planning was in some respects the exact opposite from the garden city ideal. Architecture was seen as a public spectacle, to be designed on the basis of scientific principles. Le Corbusier, its main proponent, wished to clear away what he saw as all the chaos and confusion of the past, and remove the clutter of slum districts and start again. He saw individuality, and little individual private houses, as detracting from the efficient design of modern society, and said, 'we must create the mass production spirit' (Casson, 1978). Plato's guardians were not meant to be corrupted by the sentimentality of family life or private property ownership, but in the modern socialist state nor were the workers. Le Corbusier envisaged cities being piled up off the ground in the form of mile-high skyscrapers, thus freeing the ground to create one huge area of public open space. This approach might be seen as the ultimate in the public side of the public/private dichotomy taking over the private realm. However, he recognized the importance of some functions carried out in the private realm, and sought to rationalize them by means of public provision. In *L'Unité d'habitation*, near Marseilles, an apartment complex he built in 1947, he included a nursery on the roof (functional?) and other communal facilities. But such ideas are more reminiscent of Fourier's regimented creations than the tradition of French urban feminist ideas going back to *La Union des Femmes* of the Paris Commune in 1871 (Thomas, 1966).

The modern architectural movement has seldom successfully created the mass-production spirit in the workplace either. People who work in high-rise offices frequently suffer headaches and sinus problems. This is conveniently blamed, impersonally, on the so-called 'sick building syndrome', rather than more accurately on 'mad architect syndrome'. See-sawing: when Richard Rodgers made a visit to the high-tech Lloyds Building in London which he had designed, I was told by someone who was there that all the office workers thumped on their desks to show their dissatisfaction.

The new post-modernist, neo-classical and neo-vernacular buildings (so beloved by Prince Charles) are often portrayed as having a more human scale and therefore being more attractive to women. Many women would find this argument academic, because, as with all buildings, once one is working or living inside it, it is what it looks like and how it functions internally which is of greater importance. The modern movement was based on the principle 'form follows function'. As Roberts points out (1991), function was not defined with reference to women's use of the home in relation to home-making, childcare, and the reproduction of the workforce, that is to the life force itself (p.106). Women appear conspicuously absent from the modern movement, in Britain especially, apart from a few exceptional women architects such as Jane Drew and Elizabeth Scott. But, there were a few famous women architects emerging in Europe especially in Scandinavia, such as Wivi Lom. In the Bauhaus in Germany it would seem that the men tried to push women into craft areas such as interior design and furniture (Hochman, 1990). Le Corbusier was extremely antagonistic toward Eileen Grey, a pioneering woman architect of the modern movement. Peter Adam in his biography of her (1987: 310) quoting from Corbusier's *L'Architecture d'Aujourd'Hui* reflects on how the great architect defaced the walls of her brilliantly designed house in France with his 'paintings' (graffiti) because he could not bear to think that a mere woman was more creative than he. In fact some would argue that Corbusier stole many of his ideas from her, but distorted them so that the functional principles of the modern movement because less woman-friendly in the process (Adam, 1987 and circulating tales).

Jane Drew was later associated with Le Corbusier's development of Chandigargh in the Punjab. Because of purdah requirements the *machine à habiter* approach had to be modified to provide a screened roof area in some of the housing as part of the *zenana* (separate women's space in the house, p.80). Elizabeth Scott, who was not limited in her styles or interests solely to the modern movement, frequently appeared in the journals. One would be forgiven for imagining she was the only woman architect in Britain! She is famous for building the Shakespeare Memorial Theatre at Stratford-upon-Avon (1927–32). She won the commission in an architectural competition 'judged blind' (no names). McCallum (1946: 9) comments that the (then)

Prince of Wales opened the new theatre at Stratford-upon-Avon 'to find that it had been designed by a woman' and then went on to bemoan the lack of women architects 'in spite of ten years of women architects wanted advertisements' (I found no evidence of this). She was a member of the Fawcett Society, involved in building the Millicent Fawcett Hall. See-sawing to present realities: in 1992 this building was threatened with demolition by Westminster City Council, and was eventually saved, after much persuasion, by a listing at the last moment by the Department of the Environment. It is the only purpose-built building associated with the suffrage movement (*Guardian*, 11 February 1992: women's page).

In America, especially in New York, the high-rise architecture epitomized the thrusting style of the New World. See-sawing: as a student, the symbolism of the modern movement was a blind spot to me. One architecture lecturer was always going on about the phallic powers of high-rise buildings and I really thought he was referring to the building material used, imagining he was referring to some version of ferrous concrete. Le Corbusier had visited New York, his main claim to fame not being inventing the new architecture, so much as making it popular and respectable in Europe where traditional, eclectic, decorative architecture was the only acceptable style for public buildings. In Italy too, such architects as Sant'Elia had developed Futurism. This symbolized the triumph of man's science and technology over nature, incorporating all the maleness that goes with sci-fi utopias. Ironically, in New York, which had the tallest modern buildings in the world, the only men brave enough to risk their lives to build these modern structures were native Americans (Hawkes, 1991: 87). Members of the Mohawk and the Iroquois peoples to this day bravely specialize in high-rise construction, because of the ancient, spiritual faith their tribal heritage gives them 'to walk on the high places of the earth'.

The great men of modern architecture were not as pristine and pure as the style of their buildings. Frank Lloyd Wright was famed for his creation of Autopia, that is Broad Acre City, designed around the motor car, based entirely on on one-acre homestead plots and a way of life centred on the ideal of the family dependent on the protective husband. But he was implicated in fires and murders linked to his love affair with a client's wife, and some say he 'borrowed' the drawings he used to get his first commission (cf. Fischman, 1977). Le Corbusier is associated with a range of shadowy wives (such as Yvonne in the 1930s, Adam, 1987), women and mistresses (Adam, 1987). Many women planners would argue that his lack of knowledge of an ordinary home life is relevant, because it explains how he could possibly imagine that a house could be a machine for living in (Ravetz, 1980: 43). Many of the great men seemed to have little contact with ordinary people. Geddes spent much of his time living and working up a tower in Scotland, and surviving money from a female relative (a popular strategy, while a man waits to become recognized as a genius) (Boardman, 1978). It is important to mention these aspects of the private lives of great public figures because their personal attitudes towards women surely influenced the way they perceived them in relation to the course and content of their public professional work. Likewise the words/deeds paradox (p. 24) is evident in the lives of modern famous planners who say publicly they believe in equal opportunities, but who manifestly hold traditional attitudes towards women in their private lives.

8

PRODUCTION AND CONSUMPTION
Post-war planning

IMAGES OF POST-WAR REALITY

Producing paradise

Many members will remember that whenever Raymond Unwin, our beloved second president, addressed us he was accompanied by his wife who used to sit, serenely knitting, listening with rapt attention to every word that fell from her famous husband's lips. Her presence added grace to our meeting, and we felt the warm stimulus it gave our lecturer.

(The Planner Vol. 35, No. 4: 129, April 1949,
Obituary for Lady Unwin)

Little would one imagine from this statement that women had had a primary role in the development of modern town planning (Greed, 1992c). After marking time in the inter-war period, town planning took on a power role within the context of post-war reconstruction under patronage of the Labour government of 1945. But, as Elizabeth Wilson put it, post-war reconstruction only brought women 'half way to Paradise' (1980). The Welfare State was culturally constructed by Beveridge around the traditional roles of the man as breadwinner and sportsman, and the woman as housewife and mother, not as worker or zone zapper. As after the First World War, planning was linked to public housing programmes (Hall, 1989: 98), but a new development plan system introduced in 1947 gave planners increased powers to control all types of development, and to imprint their gendered beliefs on the built environment more comprehensively. Although the planning profession prioritized land-use issues, it took a greater interest in aspatial issues related to economic development and social policy as a part of the apparatus of the Welfare State (Lake, 1941). Planning prioritized policies related to production, which was socially constructed to emphasize industrial 'male' forms of work. This marginalized all types of work and property development that might be seen as coming under the umbrella of consumption, such as shopping development, to the detriment of women. Jean Mann (1962: 170), a Labour MP, warned the party in the inter-war period that it was ignoring consumption at its peril in formulating regional policy. She argued that the needs of house-wives as well as miners (consumers as well as producers) should be considered in debates over the price of coal.

Heaviness or levity

The 1945 Distribution of Industry Act sought to create an efficient balance between the distribution of work and workers regionally. Some types of industry were considered more worthy of attention than others. This would seem to be based upon the degree of maleness, heavy industry being particularly valued. An article in *The Planner* (Vol. 30, No.5: 171–4, July/August, 1944: all references are to *The Planner* unless indicated) about 'planning and industry', actually defined the 'heaviness' of industry relative to the proportion of women in it as one of the criteria. If women were involved in heavy, primary extractive industry, they were more likely to be the subject of sexual innuendo than honour. Gibbon (1942; 116) commented, 'in the days of my childhood, young women were still employed at the pithead, and fine strapping wenches they were with tongues as sharp as old London busmen'. Adshead (1923) saw the cotton and wool industries as light women's work. Similar gendered assump-tions about the importance of women's work are found in Christaller (1933). McCallum (1946: 107) defines primary industry as of national importance, whilst secondary industry is seen as providing for local needs, and includes laundering and baking. Various studies have shown that female manufacturing workers are more productive per capita than male, heavy industry workers (Filtzer, 1993) and are thus of more primary value to the economy.

The government discouraged married women from exercising personal mobility of labour through patriarchal social security policies. There was ambiguity as to the place of women's work. In March 1962, in an article, 'The economic importance of commuters to the area of residence: a case study of the South Wales Coalfield' (Vol. 48: 73), Graham Humphreys showed a flow line map (p. 75) of 'the dependency of the mining areas on dormitory factors'. He found, to his embarrassment, that a higher percentage of women than men commuted to the mining areas to work because of 'ancillary' employment, so he stated that women's journeys to work was not to be taken quite as seriously as that of the men by the planners. Planners remained confused as to the importance of women's work. In 1969 C.M. Law, a geographer writing about changing female activity rates in the North West of England (Vol. 55: 396), stated, 'whilst the men have little choice but to work, most women usually have full-time domestic duties for at least part of their life'. Many women did and do both. Such items also show the growing influence that geographers were having on the profession, both as external philosopher kings, and as recruits to local planning offices. As a student in the late 1960s I was bewildered by the maleness of employment issues presented in lectures, and kept wondering where all the millions of women who worked in offices and factories were in their theories. Living in South London I had seen such women going to work on the buses everyday all around me. But such bias about women's work was enduring. It was even seen as a bit of a joke. Even in the 1970s, *Planning* (No.46: 2, 11 January 1974) states 'Dr. Eversley will be exhausted after his investigation of female activity rates. . .. Mr. Muggeridge will write to him in disgust'.

New towns for the old boys

New towns employment policy was ambivalent about women workers. Many employers welcomed them as a source of cheap part-time labour, but employees often considered them a likely threat to male jobs and wage levels. In July 1956 Thomas Bennett described how male workers were cushioned from such challenges, explaining in his article, 'Progress Report on the New Towns', (Vol. 42: 176) in respect of Crawley, that it 'has been possible to exercise considerable choice in employment structure of the firms involved, ... [achieving] two thirds male highly skilled and a minimum of female employment and a minimum of unskilled labour'. In contrast, J.M. Jackson in 'Dispersal, Success or Failure?', in January 1959 (Vol. 45: 47) discussed 'subsidary wage earners', who were defined as housewives, apparently seeing this reserve labour force as an essential asset in overspill developments. Whilst women are personally condemned for not being 'serious workers' if they are only part-time, it would seem they are seen (as now) as essential in providing 'flexibility' for the employer. But there was no comparable flexibility in land-use zoning to enable them to combine their two roles. A letter in January 1962 (a harbinger) (Vol.48: 21) from K. Cotton recommended the value of mixing industrial and residential uses. Pragma (the editor) replied 'yes' provided effective control over vehicles was maintained: a new reason for zoning. With increased traffic, segregatory zoning of work and home could now be justified as a means of keeping cars and lorries out of residential areas. Within this correspondence the question of crèches is raised, with the statement, 'the industrial use does provide a service to its neighbourhood, because women can be employed at unorthodox times with crèches provided for children'.

COMMERCE AND CONSUMPTION

Office development

The growth in service-sector employment, such as in offices, shops and small businesses, was, relatively speaking, not a primary concern of post-war planners. The white-collar sector of employment had expanded throughout the first part of the century, and was becoming increasingly feminized. Women's office employment was becoming one of the largest sectors of employment in post-war Britain (Crompton and Sanderson, 1990). Such women were, in a sense, zone zappers working in the public realm, but were carrying out a 'private' gendered servant role therein, as assistants and secretaries. The old left might not have seen office jobs as real jobs although they provided essential employment opportunities for women (Bruegal, 1983). Part-time workers were not generally seen as real workers. I see them as pioneering zone zappers, see-sawing between public and private realms, seeking to negotiate and transcend such dichotomies in their daily lives: albeit out of financial necessity. Women's employment was not necessarily counted in economic surveys (WGSG, 1984: chapter 4). The economic growth of London and the South East was often presented in a negative light because it was seen as drawing resources and people away from depressed areas. The region's prosperity might have been seen as being based on parasitic tertiary economic activities, that is, on the provision of services rather than the extraction of raw materials or the production of goods by pure, male, manual labour. Office development was likely to be seen as evil (on the dirty side of the clean/unclean dichotomy) being a manifestation of corporate capitalism, and the new middle-class workers might be seen as lackeys of the bourgeoisie (Bruegal, 1983). But public-sector bodies were as much generators of office employment as private firms, although possibly at that time the work force in the former was more male than in the latter. From a socialist viewpoint, such forms of employment might be seen as tainted because they were serving the needs of an economy based on consumerism, materialism, and *rentier* financial activity. In contrast, working-class men who worked for multi-national capitalist organizations were seen as oppressed victims, not as collaborators.

One would have thought the planners would have developed clear policies on commercial development, because they apparently condemned it so much. But, far from applying strong controls on town-centre redevelopment, some local authorities went into partnership schemes with private developers to facilitate reconstruction schemes and in doing so laid the foundations of the high-rise commercial property boom which was to occur in the 1960s (Marriot, 1967). Commercial developments, which were in need of control and pro-active planning were not central to the field of vision of the planners, as their interests lay elsewhere: with industry in planning for a mythical, municipalized working-class. Even in the 1960s, at the height of the property boom (Marriott, 1967) one is overwhelmed by images in *The Planner*, of working-class men, both in photographs (of men with cloth caps engrossed in heavy industry) and in terminology (for example in Vol. 52: 319, 1966, 'A method of forecasting employment', by Thomas Beynon). The explanation may be found in the dichotomized belief system held by many planners. White-collar work, especially that undertaken by women, fell down the gap in traditional divisions, between capital/labour, bosses/workers, mind/muscle and brain/brawn, occupying an ambiguous middle ground.

The lack of appreciation of the importance of office development may have been to the advantage of higher patriarchal, capitalist interests. As stated, men are not a unitary group, and patriarchy often works in mysterious ways its wonders to perform (Greed, 1992a). Planners are just one fraction in the property fraternity, and not necessarily the most powerful group. Also, because of the lack of qualified planners, it was members of the surveying profession, predominantly male and commercially orientated, who did much of the planning work in the immediate post-war period (Greed, 1991: chapter 5), and this must have contributed to the non-inteference stance towards much commercial development. In fairness there was a small minority of planners who were aware of the importance and nature of women's work. An early post-war article by F.A. Menzler (Vol. 38, No.5: 116, March 1951) giving 'an estimate of the daytime population of London', was ahead of its time in including part-time and service sector workers. The ambivalence towards controlling business may also partly be accounted for by the changing political climate, which generally moved to the right in the 1950s. But the strong planning system was not dismantled and other types of private sector developments were controlled. It was within the powers of the planners to introduce greater controls on the commercial sector if they had so chosen.

So office development went ahead, and high land values were retained, indeed enhanced in central areas, by the perpetuation of commercial property locations: in spite of the formidable planning system. Eventually during the 1960s planning policy controls were introduced which sought to decentralize and restrict office development. Unable to break the mould, planners again applied quantitative density controls (such as Floor Space Index). William Holford proposed the development of a pedestrianized precinct around St Paul's Cathedral in the City of London (Vol. 42: 214, September–October 1956,) relating the discussion to issues of floorspace density, not anticipating what the future Prince of Wales (then 8) would say about the scheme 35 years later. At the time it was seen as a gem of urban renewal (cf. Vol. 49: 344, 1963). The idea of requiring the provision of crèches for the convenience of women workers was not even on the agenda. Policies of restriction and decentralization not only reduced employment opportunities for women (WGSG, 1984; chapter 4), but limited economic options in inner city areas as a whole, arguably contributed in part to inner-city decline.

Retail development

In a planning system which prioritized production, there was little space for shopping, which is a prominent manifestation of consumption, an activity related to housework, yet an essential food gathering activity (Bowlby, 1989). Whilst retail development might be tolerated, as discussed above, as a type of commercial property, the activity of buying and selling, that is trade, and shopping still might be seen as a somewhat unworthy activity falling on the wrong side of the dichotomies (p.12). The female activity of shopping was often characterized by the planners as a somewhat frivolous, selfish pastime. Shopkeepers were portrayed as parasitic beings and, like shop assistants, not real workers (Webb, 1990), whilst women shoppers were seen as wasteful. H. Myles Wright in his town planning textbook (1948: 50) stated that whilst men needed 19/- a week to keep them in food, women needed a shilling less, and also commented on women's 'inefficiency in purchasing'. This is a classic example of a public figure passing judgment on what might be conceived as a private domestic activity which he was unlikely to have been responsible for himself. These attitudes continue well into the present, resulting in an under-appreciation of the value of the service and retail sector and an under-provision of suitable retail units on housing estates.

There seemed to be a considerable condemnation of self-employment and small businesses from planners for many years Over the years I have often heard derision and contempt expressed for both shopkeepers and shoppers (and housewives) from individual planners in lectures, conversations and throw-away comments. Part of the reason might be that local authority planners often resent what they see as ignorant shopkeepers being elected as councillors and thus getting on planning committees. Ironically, in the 1970s and 1980s the new left reconceptualized small businesses as a means of achieving equal opportunity and independence, and of providing essential local retail outlets, rather than portraying them as manifestations of incipient capitalism, and help was offered to small businesses. For example, GLEB (Greater London Enterprise Board) funded projects run by women and ethnic minorities. As stated in chapter 4, for outsider groups (especially new immigrants) private enterprise, such as starting up a shop or a business, has often been the only way to earn a living.

Unlike the planners, surveyors serving the private sector were only too pleased to meet the needs of the retailer. However, there was a catch. They were likely to define the user as the developer, retailer, tenant and distributor, *not* the shopper herself (Greed, 1991: 79); and chiefly serve the needs of big businesses who could pay for their services. The agenda seemed to be dictated by the commercial property fraternity, and by a practical need to plan for the motorcar (Tetlow and Goss, 1968). However, few women used cars to shop in those days. There is a strong input from surveyor-planners, for example an article by David Overton, 'Planning for Shops' (Vol. 41: 9, December 1954) which is all about the functions of a shopping centre from a retailer's viewpoint. It is chiefly concerned with expenditure, not with the facilities or choices offered to the shopper (q.v. p.45 on triggers). In local centres planners seemed more concerned with archaic change-of-use regulations inherited from Victorian public health legislation, as to whether a shop sold live maggots or tripe, or whether petty shopfront and advertisement controls were adhered to, than whether a development was well designed and accessible for users, with a good mix of food shops.

Shoppers complained about the unfriendliness of the modern architecture, and the impractical design of the new post-war shopping centres, such as the excessive numbers of steps, and unnecessary changes in level, wind tunnels, lack of public conveniences, and inadequate sitting areas and meeting places. Many of the post-war reconstruction shopping precincts were designed with more than a backward glance at grand manner, traditional planning – of planner as designer – with the setting out of wide avenues and windy precincts (Plymouth and Coventry respectively). The newly redeveloped Elephant and Castle site in London (Vol. 42: 161, 1956) was praised: a scheme which has been the subject of much public criticism ever after (and several face-lifts). However, some planners were doing their best to create less inhuman environments. Hugh Wilson (Vol. 44: 194, July 1958) in an article on 'Civic design and shopping centres' says 'no building should be more than 200 yds or 3 minutes walk from bus or car', and even suggested radiant heaters under canopies for pavement cafes (*ibid.*, p. 197) to create a continental outdoor living room touch! Meanwhile many existing local shops and even whole high streets were being demolished in the name of slum clearance or redevelopment. Where new precinct shopping development, or enclosed mall schemes, occurred within central areas it often blighted existing businesses and reduced shopper choice.

Planners were soon dreaming of out-of-town shopping centres reached only by car. J. Seymour Harris (Vol. 47: 243, September 1961) in his article 'The design of shopping centres', with reference to Canada describes the 'artisan' in his car doing his Friday night shopping. To his credit, he is the only planner, so far in the journal, to mention the need for public conveniences in shopping centres (or anywhere), which he says is 'a need often disregarded on account of cost, but that financial assistance is sometimes possible from the local authority'. This was necessary if people were going to have to travel North American type distances to reach these centres. For example, in Toronto I can attest that washrooms (lavatories) in shopping malls are plentifully provided, but by private developers not by public authorities as in Britain. When these are closed outside business hours there are few public facilities, particularly in small towns, (as discussed in the *Toronto Star* in the article, 'Why don't travel guides list the best washrooms', p.E1, 30 July 1993). Also Harris recommended integral play centres for children with 'a compound controlled by an attendant as in Toronto'. Thirty years later we have ended up with the distant shopping centres, without any of these facilities, and the destruction of the possibility of using local shopping centres within walking distance. I have seen out-of-town hypermarkets which are actually adjacent to housing estates designed in such a way that it is impossible for pedestrians to reach them.

Where were the women?

There were few women in planning offices to present an 'other' viewpoint. Women who had 'helped' produce plans during the war, were sent back home, or to resume their previous life as tracers (draftspersons). Colin Ward (1987) comments such a woman 'was resented because she knew too much about other people's jobs'. The only mention of women in the Greater London Plan is indirect to 'a Londoner and his family' (Barnes, 1989), but in the Foreword, six women technicians are mentioned by name. A Miss Scoffield, the secretary, is praised because she 'combined in her single person a whole staff: under her efficient and rapid hand, order has prevailed, and mountains of work have been surmounted as by magic' (Abercrombie, 1945: v) (more fool her!). Elderly male planners have told me that in the 1940s the planners who had risen within local government during the war were often 'those with flat feet', men who had been exempted from military service. Such were generally distrusted, seen either as pacifists, or as being in the pocket of councillors and unlikely to implement radical planning policy (compare 'draft dodgers' in surveying, Greed, 1991: 68). War veterans were welcomed, imbuing the planning establishment with a military ethos. A few women architects who had escaped Nazi Germany entered British planning, for example Marianne Walter who become Sheffield's first woman planner in 1946 (Walter, 1985, 1992: 253) and designed and built The Dronfield Pioneer housing scheme in Sheffield. Women are seldom mentioned in post-war plans, and are then indirectly referred to as housewives (Roberts, 1988a and b, Morris, 1986). A woman's viewpoint was not necessarily valued in those days. An article written by Annette Seale on 'the plan for a small town: a practical approach', was much criticized (unfairly in my view) by A. Thorburn and A. Swindall who wrote 'she summarises the difficulties . . . she does not attempt to suggest detailed solutions' (January 1961: 19).

The planners, 'just didn't think' about the consequences of their plans. There is a photograph of a new Arndale enclosed shopping centre (Vol. 60: 747, 1974) which includes within its frame a glimpse of a shop window full of prams, just at the time when the combination of poor shopping centre design and the necessity of using transport to reach such centres made it increasingly difficult to take a pram shopping. Planners' dreams of the future were equally unrealistic. An article by R. Davies and D. Edyvean waxes lyrical on teleshopping by computer link (Vol. 70: 8, August 1984) with a photograph of an elderly woman apparently doing so. In the real world the number of the schemes (such as STORX) are negilible. It is not profitable to take, and deliver lots of little orders for old age pensioners in their homes when retailers can attract big spenders in cars who come to them.

Articles written by women on shopping are rare. Those who did have the opportunity to write about shopping in the journal adopt a somewhat commercial viewpoint, as in the case of an article by Margaret MacKeith (Vol. 71: 9, January 1985) (who received such a bad review of her book on shopping centres, p.103). Other women adopt the men's agenda and discuss issues such as commercial deliveries and goods traffic (Pain, 1967). In 1974, Trevor MacMurray addressed the question of 'planning and the market' (Vol. 60: 826) but planners, generally, seemed to live in a different world from surveyors and property developers. Thus they were not adequately involved in the very subjects, such as retail and office development, which directly affect women. Another theophany was the announcement in 1972 that Heather McCrae had set up as a private consultant (Vol. 58: 194, 1972), the beginning of a trend for women to set up in private practice (such as Yvonne Phillips today). See-sawing an RTPI retail working party which reported in 1988 did not appear to prioritize women's issues (Vol. 74: 9 October).

CONTROLLING URBAN SPACE

I will now consider three manifestations of town planning, namely transportation planning, public open space planning, and urban systems theory. All three were concerned with the organization of urban space upon a grand city wide scale on the basis of gendered perspectives.

Transportation planning

With the growth of the suburbs, decentralization, lower residential densities, and segregation of land uses through zoning, transport became an increasing problem for many people. Prior to the war, it was apparently assumed that women had the time and inclination to walk everywhere, the journal stating in May 1939 (Vol. 25: No. 7: 236), 'the housewife likes a little competition and shops half a mile from each other'. Planners were beginning to address the problems of commuting, which appeared to be an entirely male activity undertaken by motorcar (McCallum, 1946: 183). Women encountered some opposition from the beginning in using cars themselves. For example, in 1904 the Ladies Automobile Club was founded because the AA (Automobile Association) barred women from membership, but subsequently relented in 1905.

In 1928 5,489 people were killed on the roads, but apparently buses killed many people as well as cars. The police rather than the planners in Britain appear to have taken the initiative in transport planning. Alker Tripp a police commissioner (1942) wrote a seminal book on transport planning that incorporated many American ideas. These included the concepts of the precinct and the Radburn neighbourhood in which pedestrians were segregated from motor vehicles by means of separate footpaths and cul-de-sac. This idea was much criticized by women for increasing the likelihood of mugging on deserted paths and underpasses, although, possibly, reducing road fatalities. It reflects the propensity of planners to try to solve problems by segregating and dividing the components involved, i.e. zoning, usually to the advantage of the 'male' elements, in this case, prioritizing cars over people. Adshead (1932: 81) stated that the width of footpaths should not be less than 5 feet to allow the passing of two perambulators, thus signalling 'footpaths are female'.

There was a need to control and plan for traffic. A black American engineer,

Garrett A. Morgan (1887–1963), in February 1927, invented the world's first electric traffic light in Cleveland, Ohio. He combined many interests being an early transportation expert; a prolific inventor across a range of fields; and a civil rights activist (12 November 1991: 4, *Guardian Education Supplement*), and one of the few black people in planning. See-sawing: I found only one Afro-Caribbean, Black, British full-time lecturer teaching town planning, in any college in 1992, namely Patrick Loftman at Birmingham (University of Central England). There may be others apart from him but they are hard to find.

Transportation planning was not considered to be a major component of post-war reconstruction planning. In January 1955 (Vol. 41, No.2: 49) Colin Buchanan wrote about car parking in central areas, and subsequently produced the prophetic book *Mixed Blessing* (1958) on planning for the motorcar, which has some odd comments in it. A photograph of a large car parked over a pavement, with a passing businessman casting an admiring glance at 'her' glossy chassis as he reverentially sidles past, is captioned 'The pedestrian cheerfully suffers the loss of his amenities' (Plate XXI). In November 1955, (Vol. 41: 294) Henry Dawson wrote on 'The costs of street congestion' and was ahead of his time in stressing the time-budgeting implications for passengers (Bhride, 1987) but went on to discuss 'the erratic behaviour of livestock' as delaying traffic flow! The influence of North American traffic planning was beginning to be felt, for example in September 1956 (Vol. 42: 228) elevated roads were discussed. By the 1960s all manner of precincts; traffic architecture (Vol. 57: 3, 1971); and environmental areas, with segregation of pedestrians from vehicles became popular, reflecting beliefs about the separation of public/private realms. Buchanan the traffic engineer became President of the Institute (Buchanan, 1963). Traffic architecture that drives pedestrians underground in subways and pedestrian precincts increases the likelihood of attack for women. Many women pedestrians prefer traditional main roads with bus stops.

Some planners believed traffic problems were caused by the selfish working-classes having it too easy. In 1958 H. Myles Wright blamed the new 9–5 day for the rush hour (Vol. 44: 182). The introduction of the five-day week was seen as increasing the pressures in the countryside caused by cycling and camping (not cars yet) (Vol. 45: 5, December 1958). In the 1950s 'the reasonable man' was still referred to as 'the man on the 38 bus' (Vol. 45: 161, 1958). The average commuter was seen as male. In May 1960 (Vol. 46: 135) the journal showed a set of illustrations entitled 'Visible Changes in Cities', by David Thomas, showing pedestrian man, equestrian man, and automotive man. The journal advised (Vol. 46: 230, October 1960) 'think big and provide an additional parking space for the second car'. In the 1960s transportation planning was a new area of expertise, and therefore (as in the early days of computers) it was unclaimed territory and more open to women. Betty Yule wrote an article on a traffic survey in Cambridge she undertook (Vol. 43: 116,

April 1957). Likewise, Barbara Castle was not only appointed Minister of Transport in the 1960s, but was the first woman to write a non-housing article in the surveyors' journal (*Chartered Surveyor*, Vol. 98, No.10: 516, April 1966). Sylvia Crowe made a career out of road landscaping (Vol. 47: 49, February, 1961) writing, *Landscape of Roads* in 1960, combining a woman's traditional 'prettifying role' (Greed, 1988: 138) with an interest in this new area of planning. Other women were taking on another customary role for women in planning: the protester and community activist. Jessica Albery (associate member) advocated rail against cars in a letter to the journal (Vol. 46: 274, November 1960). Transportation planning was socially constructed to incorporate planners' favourite things such as quantification, division and zoning, objectivity and impersonality. This is epitomized visually by those soulless Letraset transfers, depicting 'typical' people (male executives, dolly birds in mini-skirts), with which planners used to garnish their schemes; see for example Vol. 51: 71, 1965, showing the Letraset folk tackling an elevated walkway. The division between car drivers and pedestrians equates nicely with division between shoppers and workers, and thus between women and men. In February 1968 (Vol. 54: 61) there is a photograph of three mothers about to cross a primary distributor road with their children, captioned: 'just up the road there is an underpass'. It looks quite a way up, and the footpath does not link directly to it, but the women's problems are blamed on their perceived stupidity not the men's poor planning.

However, generally the shopper was assumed to be a non-gendered individual in a motorcar. This resulted in a craze for mathematical modelling of retail gravity attraction, based on the quantitative factors of the amount of floorspace divided by the travelling distance by car (Roberts, 1974, a mathematical woman planner (p.167)). To understand why some schemes were really more attractive than others, there is a need to see-saw into the private realm of people's daily experience of going shopping. It is vital to consider what the bus services are like, whether there are adequate facilities for women, or whether a certain chain store or supermarket has a bad reputation. It is to his credit that Cecil Clutton (Vol. 55: 85, 1969) planner and notable personality in the world of chartered surveyors wrote a letter to the journal on what he called 'the silliness of retail gravity models': and he was a commercial, not even a social planner. When I entered college in the late 1960s the computer-driven approach to planning was at its height, with all sorts of mathematical models being developed to assist the planners such as cost-benefit analysis, which marginalized social and qualitative issues. I remember being told as a student: 'if you can't count it, it doesn't count'.

Public open space or closed private space

Another important component in the city of man is 'public' open space. The conceptualization of open space has been gendered from the start, strongly linked to the male importance of the concept of 'play' (and thus to sport) which along with work, and home, forms the third component of the trichotomy identified by Geddes (p.117). Again, as Cockburn observes, 'to be male is to occupy space' (1985: 213) and it is noticeable as one looks at the evolving demarcation of space through planning that men increasingly give themselves large, unbounded areas of space within the 'public' realm. In contrast, private space, within the residential neighbourhood, and within the home is increasingly restricted. The standard size of council houses became smaller in the post-war period. Throughout the century the open areas which had been provided for women were often heavily loaded with folkloric, homely, images of peaceful, bucolic life. Open space often seemed to be imbued with almost magical powers to create a harmonious society.

Both the garden cities and the new towns built in the post-war period were based, to varying degrees, upon the neighbourhood unit. These, usually containing around 5,000 people, constituted a community area provided with its own local shops, schools and amenities. A central area of open space was an important component of the neighbourhood, as in the case of the green heart of the Radburn super block (Greed, 1993a: 201). This space, a descendent of the village green, was popularized by the pioneers of the garden city movement. It was not only meant to provide a safe play area for frolicking children, but could also apparently create a sense of community which transcended class conflicts. Creese (1967) commented in the 1920s that if 'the squire's son and the agricultural labourer's sons play together on the village green' then, to paraphrase, everyone would grow up happy and classless. In the same rustic vein maypoles were erected in post-war new-town precincts for residents to dance around to engender community spirit. People would oblige if it guaranteed their being seen as good tenants! Open space, especially green grass, appeared to be seen as the panacea for all social and political problems for many years to come. The National Playing Fields Association thrived in the inter-war period, recommending 6 acres of playing fields per thousand population. To create this standard women are roped in, numerically, to support the need for spaces they were unlikely to be welcomed to use. This organization has been linked to town planning from the beginning, and it is a statutory requirement that it is consulted on a range of planning matters (Ashworth, 1968: 229). An article on the need for school playing fields for 'children' (Vol. 16: 188, July 1929) was clearly about boys. Professor Adshead (1923: 52) in his foundational textbook on town planning had commented, 'games like net-ball, hockey and lacrosse can be played on the area provided for cricket and football, when these are not used'. He adds, 'immediately outside the clubrooms there might be laid out a garden, and in summertime tea could be provided amidst the trees and flowers, the meeting hall would be reserved entirely for political and social meetings and lectures'.

By the 1960s some planners, in North America at least, were questioning the holiest space of all, the playing field. In March 1960 (Vol. 46: 98) there is a review of *Playgrounds, and Recreational Spaces* by Alfred Lederman and Alfred Trachsel (1959), a book which prefigures what Jane Jacobs was to say on 'open space' (1964). The authors criticize what they call the 'tedious acres' of playing fields loved by the planners, favouring instead all-age specialist space provision. But the reviewer poured scorn on the authors' ideas of parks with duck ponds. An article entitled, 'How much urban space do we need?' included a photo of men playing rugger ('we'), and discussed male winter, field, team sports (Vol. 53: 144, 1967). The reasons for the 'green grass fetish', and the idea that 'green is good' may relate to ancient beliefs about fertility. Grass is a symbol of the seed of man, and thus of fruitfulness and population. But it is also a symbol of grazing and thus of agricultural settlement, and so it represents mastery over nature and dominance over space (cf. Ardener, 1981). An area of green open space may also fulfil an important 'religious' function as a neutral zone, a holy area apart. Perhaps modern man cannot stand the tensions created by his division of the world into home/work, public/private zones. He needs somewhere set aside separate from these dichotomies, in which to escape: a wild zone, for relaxation through play, in sport. Zones can also be created by music, to define social space or age bands. One of MTV's programmes, significantly, is called the party zone. Computer games are frequently structured around zones of escapism, whence comes the phrase 'zone zapping'. But, women who zone zap, and take authority over male space where they should not be, might themselves be 'zapped' by military micro-waves, as at Greenham Common, apparently.

A hint of the origins of open space as a neutral zone, where salvation is found, is given in an untypical debate in the letters pages in May 1956 about the definition of the *migrash* in the Old Testament (Vol. 42, No. 6: 129). Leslie Ginsberg, a planner, identified it as an early form of green belt, by reference to Numbers 35: 2 and Isaiah 5: 8. He argued that it was an open space, not a suburb as translated in the King James version. Another planner replied, quoting Ezekiel 45: 2, that the *migrash* could also be interpreted as a city of refuge, a sanctuary, and thus a satellite garden city. But, arguably, a *migrash* is a suburb, not to be confused with the six cities of refuge in Numbers 35 (Wigram, 1980: 663). I am intrigued by the idea of a separate escape zone where one can step out of the world, as it is a theme developed by Kessler in discussing the wild zone of feminist utopian imagination (1984: 19). Whilst for many men it is an external public space for team sport, for many women, relatively speaking, it may be an internal space: a 'secret garden'. Compare also the Jewish concept of the *eruv* (c.f. *Planning Week*, Vol. 1, No. 14: 13, 9 December 1993), a zone where Sabbath laws are lifted.

Whilst there seemed to be little restriction on the provision of public open space, a far tighter control was kept on space within the private realm of the housing estate. Before the war, it was commented approvingly in the journal that the 'young mother' likes to have her garden (Vol. 21, No.6: 164, April 1935) but in the immediate post-war period criticism of the low-density garden city approach inspired the question, 'Is town planning merely gardening?' (Vol. 38, No.10: 253, September–October, 1952). Questions were raised about the value of each house having a garden. The appearance of rows of small gardens each surrounded by fencing was compared to pens in a cattle market (Vol. 25, No. 9/10: 289, 1939: Frontispiece). Major debates about the utility of the gardens were entered into, it being argued that land lost to residential development from agriculture was more intensively cultivated as gardens raising food production (Vol. 38, No.3: 54, January 1952) – provided everyone grew vegetables. Other uses for childcare, washing and storage, and the importance of having a private exterior zone, were not mentioned. One senses a subtext beneath such debates that the domestic realm was considered to be really rather trivial not worthy of too much space being wasted upon it, relative to sport. The 'density debate' continued, but the tide was turning against its importance as the basis of planning policy. In March 1956, Frederick Osborn (Vol. 42, No.4: 92) stated in a letter that he considered density to be a fictitious concept! A letter from Elizabeth Denby in April, 1957 (Vol. 43, No.5: 122) attacked the cliché of density, saying it was a surgeons' rather than a physicians' solution. By this time the high-rise movement was just beginning in Britain. Denby wrote to the *Manchester Guardian*, to express concern about high-rise blocks (Vol. 43, No.7: 181, June 1957). One might ask 'why' the high-rise movement caught on. It seems to me that in each generation a new 'brat pack' has to emerge in the profession which adopts ideas radically different from what has gone before, to establish its importance. It challenges the achievements of the previous generation, as it seeks to take over the running of the planning tribe, from the 'old men'. Perhaps it is not a real conflict, but the old men expect the young men to indulge in this game to prove their worthiness to take over. Differences of opinion between generations may not be as deep as they seem.

The destruction of working-class areas in the name of slum clearance under the 1957 Housing Act caused much distress. Residents had little option but to live either in distant green-field housing estates, or in the new high-rise blocks. This was top-down planning 'for' the working-classes. As Ravetz comments (1980), how would you feel if a notice came through your letter box informing you that your house was unfit (a eugenic word) for human habitation, when you had felt a pride in keeping the house decent, and being respectable, and making modest improvements? It was subsequently stated in the correspondence (Vol. 44, No.1, 1957–8: 29) that 'density is a basic mathematical concept which does not cope with the qualitative aspects of town planning'. But it was not safe 'yet' to decry density unless one was either a very high status male or an outsider female (who did not 'matter'), because the planners' power depended on it. Public opinion against high rise appeared barely to register with the planners, although they were meant to be planning on behalf of society. But a Mrs Muriel Smith (community development officer with the LCC), criticized pre-fabricated, systems-built housing schemes. She cited examples of bad internal design (Vol. 47, No.1, p.9: 75, 1961), encapsulating all the points of the insides/outsides debate as to planning's control over housing design. The arguments for high-rise development (although 'foreign') fitted well into the British density discourse, with its images of the land being swamped by 'too many little houses' which presumably took up too much space. Such discussions are redolent with eugenic and classist overtones. For example, an article by Max Lock (Vol. 46: 264, November 1960) in relation to the changing role of the Institute of Housing, asks: 'where are the new houses to go?'. 'Housing' as a subject was beginning to be recolonized by men (and angry young men at that) (Greed, 1991: 64). The Institute of Housing (for male local government officers) took over the Society of Housing Managers (for lady housing managers) in 1965.

Urban systems theory

The nature of the 1971 Act was influenced by a systems approach to planning which developed in the late 1960s (McLoughlin, 1969). It combined in one methodology the new computer-driven, mathematical approach to planning (overladen with transportation planning objectives), and an obsession with the future reflected in the use of cybernetic jargon derived from the moon-shot program in America. The planner would act as the 'helmsman' guiding the system into the blue yonder. See-sawing: I remember sitting in a tutorial group in about 1968, in which we were discussing 'the future' (as one did in those days) in which apparently 'everyone' would only work three hours a day and 'play football' for the rest of the time, because science would provide machines to do all man's work. *'But what about women?'* I asked innocently, *'a woman's work is never done'*. *'Don't be silly, we're not talking about that'*, came the humiliating reply from the lecturer. That incident marked the beginning of my journey towards urban feminism.

Treating the city as a system, a set of inter-related, impersonal, objective component land-uses that generated traffic, further distanced human beings from the urban scene, and allowed no space for the realities of everyday life. Whilst the zoning tradition in planning had always stressed the importance of division and segregation, a systems view (apparently) ignored the divisions and stressed everything was part of a greater whole. The categories which were to be interconnected remained male and the processes which linked the components were mainly 'male' public-realm activities. McLoughlin's writings are full of gendered assumptions, for example (1969: 412 *et seq.*) he states: 'the most numerous decision-makers are households. The householder or family man . . . identifies opportunities on the basis of his experience . . .'. After this subtle bit of elision to exclude 'the family woman', he discusses 'man's decision making powers'. He made planning appear unpolitical, as if there were only one 'scientific' right answer. Class and gender differences, and the power of the private property sector, were transcended through this apolitical, 'scientific' view of urban dynamics. This was not a theoretical ethos conducive to discussing women's issues. In a similar manner today, class and gender differences that were previously stressed in structuralist sociology can be made to appear unimportant, as just two interesting considerations among many issues, in post-modernist sociological thought. The 1971 Town and Country Act introduced a new system of development plans, called 'structure plans' which were meant to set overall policy goals for the future development of the city, reflecting a systems methodology. Dichotomously, the act encouraged public participation, just when scientific jargon, and abstract diagrammatic plans, dominated the planning discourse.

9

URBAN SOCIOLOGICAL
PERCEPTIONS OF WOMEN

DICHOTOMIZED PERCEPTIONS

Traditional family values

In this chapter I will rerun the spool and consider how urban sociological perceptions of women and their needs have been reflected in the planners' world view, by drawing illustrations from *The Planner*. I will also make observations drawn from my own life experience as a previous inner city resident to illustrate 'the problem', thus see-sawing from the public to the private realm. To reiterate, this should not be seen as an interruption but as a mainstream theme because personal accounts are integral components in post-modernist feminist sociology (Stanley and Wise, 1993). In the 1950s urban sociology was given a new lease of life as researchers studied the effects of slum clearance and rehousing on people's lives (for example, Young and Willmott, 1957). There were also studies of the communities in the post-war new and expanded towns (Aldridge, 1979; Stacey, 1960). Incidentally, Margaret Stacey was one of the few women sociological researchers to be published in her own right in those days. Later she became a second-wave feminist writer (Stacey and Price, 1981). However, in general, urban sociology was so highly gendered and classed that it did little for 'women and planning'. Arguably urban sociology reinforced planners' patriarchal view of society, albeit from a new aspatial, as against traditional spatial viewpoint. Up until the mid-1960s, at least, planners appeared to hold unrealistic, sentimental images of urban life, which were subsequently replaced, following a period of urban unrest, by somewhat more realistic images of the different types of people who make up society including, eventually, 'real women'.

A review in June 1961 in *The Planner* (Vol. 47: 175) of Young and Willmott's book *Family and Kinship in East London*, stated that its contents will 'strongly affect' planning decisions. There is no doubt that Young and Willmott gave some emphasis to women in their study. But their fondness for seeing woman in the role of 'Mum', as wallpaper to the main action of life, suggests they were adopting too narrow a focus. They chiefly saw women as existing within the private zone of family and community as an adjunct to working-class men. They scarcely considered their participation within the public zone of work and women's other interests outside the confines of the community. From an urban feminist perspective, many see their work on the symmetrical family (1978) as nothing more than wishful thinking (WGSG, 1984), and their research on community as 'damned' (Rose, 1993: 53) because it was heavily gender-biased. Such studies are examples of middle-class man, as the 'norm', coming from the public realm of academia to study the 'other', namely working-class woman, in the private realm of home and community. Of the same ilk is Goldthorpe's highly gendered and unrealistic study of the affluent worker (1969). In this he gives the impression that increased affluence in society as a whole can be assessed by studying the changing fortunes of working-class men alone. Such definitive studies, which are still referred to today, can make women students feel marginalized and selfish, as if it's 'just them'.

Urban social zones

The functionalist social ecology theories developed in Chicago were popular with planners, possibly because the researchers analysed the city by dividing it up into zones, reassuringly implying that major social problems only existed in the deviant area, which was quite separate from the normal city. Such theories were included in post-war planning American textbooks such as that by F. Stuart Chapin (1965; and Vol. 43, No.9: 253, September–October 1957). Zonal social divisions between areas reflected ancient beliefs as to whether the inhabitants were, variously, clean or dirty, insiders or outsiders (especially immigrants), and in a state of order or chaos. Social ecology theories were generally concerned with divisions among different male groups (especially macho-deviant groups). Delamont sees aspects of the Chicago school as anti-feminist (1992). But, some of the original Chicago researchers allowed for the possibility of women being workers as well as mothers, as reflected in the urban questionnaires used by Zorbaugh (Bulmer, 1984: 103). American ethnographic studies of urban communities were popular in the 1960s (such as Whyte, 1963) which usually presented the inner city as male and dangerous (Lawless, 1981). In contrast there were several North American studies of the suburbs in which the women were presented as bastions of respectability, but vacuous, although the housewife was a 'safe' type of woman and knew her place (Gans, 1967; cf. Friedan, 1963).

Women in the inner city were viewed in condemnatory light if they did not conform to a traditional working-class housewife image, and adhere to normal gender/race/class/sex demarcations. Wilson comments on the judgmental manner in which independent, middle-class women were viewed (1991: 77) – as aberrant zone zappers and 'out of control' – especially those who chose to live in bohemia, and in the zone of transition. This may be seen as a mixed, middle zone, where conventional dichotomies were in abeyance. It provided refuge to many. Some saw it as one of those special places touched by spiritual forces because of the needs of its inhabitants, as expressed, for example in the work of the Urban Theology Unit in Inner Sheffield (cf. Daly, 1993; a cave of Adullum, I Samuel 22: 1). It was not an approved zone, but deviant and wild. Its transitional nature was seen as a temporary dysfunction, not an ideal model for future urban social structure. In stark contrast an antiseptic social mix for new town neigbourhoods was approved, made up from respectable (as against rough) working-class people (Aldridge, 1979). Studies of deviance in the criminal area, in the genre of the Chicago school in Britain, such as by Morris were extremely moralistic towards young women, and blamed much male criminal activity on the mother's perceived lack of responsibility. In passing, the area's influence (Morris, 1958: 108) was recently recalled in 1993 with the review of the Bentley murder verdict of 1952. There was little distinction drawn between victims and aggressors in discussions of crime, or acknowledgment that the former were more likely to be female and the latter male. Little acknowledgment was given to the nature of female-specific poverty. For example, some middle-class women who had become single parents might live in the inner city because they were poor, not because they were criminal or working-class. Nor was adequate attention given to the complexity of 'other' circumstances that made men and women poor, beyond 'class', such as old age or business failure (cf. Ellis, 1991 on John Major's family's sojourn in Brixton). Indeed, poverty, race, gender and low class are often conflated and confused in such studies. Inner-city problems were seen as the fault of the residents, not their rulers (planners, teachers, social workers) who kept people down.

Popular Poplar

In post-war Britain women researchers were gradually having an impact on the content of urban sociological literature, making it more 'human'. In November 1951 (Vol. 38, No. 1: 7 *et seq.*). Ruth Glass of the Social Research Unit, Department of Town Planning, University College London, wrote an item on the new Lansbury housing redevelopment in Poplar. Several of the most senior women in planning today started off as research assistants (a traditional female role), not as planning practitioners. Also, UCL has always been one of the main sources of senior women planners. She commented (p. 11), 'a young mother from Poplar said, "I am very happy to think my baby will be born in such a beautiful place"'. No doubt the flats were much better than what the tenants had previously experienced, being developed as a 'model' housing scheme in association with the Festival of Britain initiative in the early 1950s. But I sense embarrassment and deference in the woman's response (as 'other' to the woman researcher). This 'young mother' had a husband working in the docks and two sons, and obviously did not want 'trouble from the council', so she replied to the 'lady' researcher: 'Lansbury has only one major fault that there is not enough of it'. But in the article 'truer' observations are also made about lack of local part-time work for women to help pay for the luxury of 'lino' to cover the floors.

Other articles suggested that the planners were slowly realizing that their image of woman, and her role in society, were not an accurate reflection of the real world, and her participation in the workforce. Michael Young contributed to a feature on 'The planner and the planned; the family' in May 1954, (Vol. 40, No.6: 134). This presented strange images of childcare and domestic bliss, commenting, 'old men sunning themselves on the park benches can watch the babies sprawling on the grass in front of their feet'. Innocent days, these, before the acknowledgment of the dangers of child abuse and incest in every situation; and not a nappy in sight either. On page 141 of the same feature, a Mr Cowly retorted angrily, 'Mr Young said one of the services which could be performed by the extended family was to look after children whilst the mother was out at work. If Mr Young was looking for major factors in the breakup of the family, one of them, he should have thought, was the fact that mothers of young children were out at work'. This is the period of the closure of war-time nurseries and the 'invention' of the latch key kid. But Ruth Glass countered admirably that there should be more support for mothers and less emphasis on the family. I sense a muffled, feminist debate going on beneath the surface.

Planning for people

Sociologists were beginning to question the neutrality of planning methodology, such as the 'survey, analysis, plan' approach pioneered by Geddes. James Simmie in May 1953 (Vol. 39, No.6: 126), in a seminal article on the contribution of the sociologist to town planning, approvingly quoted Catherine Bauer, who said, 'most surveys are a number of unwarranted assumptions . . . held to predetermined conclusions'. Incidentally, she (later Bauer Wurster) was one of the few American women planning theorists published in her own right (cf. Foley, 1964). While the sociologists could easily spot the silliness of the 'density debate' and spurious quantitative space standards, they missed the gender bias. This is manifested in the ethos of the journal (for example, Vol. 43: 158, June 1957, 'John Citizen and the Planners') and in attempts at public participation which naturally targeted men. Post-war planners were attracted to grass-roots planning as pioneered in the TVA scheme (Tennessee Valley Authority). Chapin contributed an article (Vol. 36, No.3: 112, February 1950) on this, praising the idea of 'fraternal public participation'. Women planners (few though they were) had been pioneers in public participation. Elsie Rogers, an architect-planner, held a public meeting in 1946 (Vol. 31, No.2: 74, January–February). It is stated with disappointment that it was chiefly attended by women, and a few men on leave, because men had not been demobbed yet. Women planners were also seen as being good at dealing with social topics. In 1942 Clough Williams Ellis wrote about the planning of an industrial settlement, saying that they wanted to know how 'real clients', that is people who had to live in the houses, would find it (Vol. 29, No.1: 40, November–December). They employed a Miss Denby to interview a selection of representative women, stating that this might be seen as a new profession for women of 'social surveyors' who might 'gain intimate knowledge of these people'.

Public participation was not seen as a key feature of the creation of post-war utopia for the working-classes. Planners seemed genuinely to believe that they knew what was best, presumably imagining their views reflected public consensus. In 1954 Lord Denning wrote an article on 'Planners and the planned: Equality and the citizen' (Vol. 40: 82, 1954), in which it was argued that the obvious legality and fairness of the system overcomes social differences as to how the new Britain should be planned. In January 1953, Lewis Silkin, the first Minister of Town Planning, (Vol. 39, No.2: 26) wrote, in a similar vein, '[planning] inspectors are the finest body of men you could find, fair minded and sometimes too gentlemanly'. Public protest had occurred only a few years earlier when Silkin had attempted to purchase compulsorily farmland for one of 'his' new towns (nicknamed Silkingrad). Some planning lawyers seem to possess an incredibly compartmentalized tunnel vision, possibly because their training makes them concentrate exclusively on the case in question, everything else being seen as irrelevant. See-sawing: I was disappointed that a certain senior judge who has a good reputation in determing cases related to planning inquiries, especially in preventing inhuman slum clearance in northern cities, in contrast, when dealing with a rape case declared that 'the level of vigour in sexual congress which was generally acceptable was higher today than in 1934', (referring to a precedent from R v. Donovan, 1934, in dealing with R v. Boyea, 6. February 1992). This is included as a classic example of the double standard of justice applied to women's private lives, even by those who are fair in their public dealings. One wonders how he would judge a 'women and planning' case.

In the new towns a more specifically social type of planning was emerging; the planners attempted to create a sense of community by neighbourhood design (Carey and Mapes, 1972). In May 1964 a summary is given of sociological studies considered to be relevant to town planning (Vol. 50, No.5: 187). Margaret Stacey's work is mentioned, but the list mainly comprised studies by men. Acceptable women sociologists seemed only vaguely aware of class, let alone gender, and often went along with the men in the creation of an image of sentimental community life (Vol. 53: 95, 1967). One cannot underestimate the influence of a woman's class background on her perceptions of others. In 1965 Gordon Cherry (Vol. 51: 252, 1965) wrote an article on the role of social planning, relating it to Newcastle, which is ironic in that much unwelcome upheaval was occurring there at the time. This was later to be described in *The Evangelistic Bureaucrat* (Davies, 1972), a study which highlights the religious sense of mission and conviction possessed by the planners at the time.

The problems of the elderly, not a popular topic in a profession obsessed with planning for the future, were being 'discovered' by the planners in the 1960s. But the problem was usually seen as being about the male elderly, who were seen as lonely and neglected (Vol. 52: 94, 1966), even though women constituted the majority of the elderly. As to women in general, David Eversley commented in the *Architect's Journal* (29 November 1967: 106–7) that the rapid spread of labour-saving devices had increased the leisure of wives and mothers despite the disappearance of servants, thanks to a combination of higher incomes and lower prices. It was believed that the new consumer goods made women's complaints invalid. Detailed research by such early second-wave feminists as Hannah Gavron proved the opposite (Gavron, 1966: Lewis, 1992a). Women were naturally seen as housewives, because that is what they were 'for', as expressed by Nick Reid in an article on Basildon New Town (Vol. 53: 297, 1967). More encouraging was an article by Eric Reade in 1968 (Vol. 54: 214) on the sociology of planning as a profession, which hinted at some of the issues which were yet to emerge. In 1969, Jim Amos stressed the importance of social planning but did not appear to know quite what it was (Vol. 55: 141). In contrast, a book by June Norris of the Midlands New Towns Society raised *The Human Aspects of Redevelopment* (1961) but this was not seen as prophetic. A patronising review is given (Vol. 54: 293, 1968), of *The Plan that Pleased* (1967). This was a personal account by Elizabeth Mitchell, (1880–1980) of East Kilbride New Town. Elizabeth Mitchell's chief pre-occupation, since the 1930s, was for the creation of new towns to relieve the congestion of Glasgow working-class homes (Lindsay, 1993: 57), albeit working in the voluntary sector and not professionally qualified. Margaret Mead (1968) wrote a paper 'Houses are Homes' for a conference at Coventry Cathedral, which suggests that religion rather than sociology provided a space in the in-between years for urban feminist thought.

DISILLUSIONMENT WITH THE GOSPEL OF PLANNING

In the early 1960s the profession entered a period of retrospection as to its achievements. The planners were losing their faith. The optimism of the post-war period gave way to a subsequent crisis of identity and confidence. The spell was broken; disbelief could no longer be suspended. A letter from a K. Hiller in April 1961 (Vol. 47: 105) identified a pattern of ideologies within the profession, prophesying that another ideological crisis was about to happen to 'the gospel of planning', referring to similar stirrings in 1940s and 1950s. A new brat pack (p. 138), product of the post-war meritocracy (Young, 1958), was about to take over. In July 1961 (Vol. 47: 183) the problem of planning atheism is identified. The journal stated, 'this heading is not to suggest that planning is some sort of religion'. Planners are then categorized into three types: true humanists, frank do-gooders, and cynical sceptics (all of which fall into my philosopher king category).

An impending split along spatial/aspatial lines in the Institute was reflected in an article by Lewis Keeble in 1961 (Vol. 47: 185) entitled 'Planning at the Crossroads'. Conflict developed between the old guard of physical land-use planners such as Keeble (1969) and the new aspatial, politically inspired social planners and sociologists. In the pre-war period most planners came from existing built environment professions. In the post-war period there was an influx of those deriving from the disciplines of geography and sociology, bringing in new spatial and aspatial ingredients. Professor Goss (1965) commented of the membership, 'are geographers some 40 times more necessary than sociologists, or 20 times as necessary as engineers and surveyors, or slightly more necessary than architects?' (cited in Marcus, 1971). Such divisions within planning are particularly counter-productive when dealing with women's planning issues, which often have both spatial and aspatial dimensions. Some of the sociological group were to espouse neo-Marxist socialism which was very different from the Morrisonian, safe, municipal version of the past: but not necessarily any more 'women-friendly'. The 1960s was the age of the angry, young, socialist man who was looking for an outlet for his social conscience and the planning profession was ripe for the taking. In retrospect, such individuals seem quaint with their sense of self-importance and righteous indignation about the ills of the working-class, often expressed in extreme rudeness to middle-class women. These men saw themselves as the intellectual elite of a generation in which less than 5 per cent of young people went to university.

On the Institute's 50th anniversary in 1963 it had just over 4,000 members (Table 2, p.195). In the mid-1960s the Institute moved to Portland Place, near the RIBA (architects headquarters) but away from the centre of power. The surveyors and the civil and structural engineers headquarters are all located in palatial buildings in Great George Street, Westminster, near the Houses of Parliament ('the men's house', Ardener, 1981: chapter 3). It is said that if a politician seeks informed opinion on a built environment issue he just pops across the square to have a drink with the surveyors in the RICS bar, and would not dream of struggling across London to the downmarket RTPI headquarters. Meanwhile, public opinion was turning against the planners. The conviction for fraud of urban politicians such as T. Dan Smith from Newcastle, previously praised in the journal for his town planning ideals, further confirmed to many that planning was not really for people at all, but a mysterious activity undertaken by members of public bodies for private gain (Vol. 51: 31, 1965). See-sawing violently: the creation of the Greater London Council in 1963 was seen by many Londoners, such as ourselves at the time, as creating yet another tier of middle-class, parasitic, paper pushers and planners, who had to be fed and housed by salaries paid for out of our rates and taxes, their existence being legitimated by a show of pretending to know how to solve our problems for us.

'The need for new ideas' was raised (Vol. 50: 6, 1964). Professor J.R. James reflected on the likely issues which would affect town planning over the next fifty years, without mentioning gender at all. In the memorable article 'where have all the wildmen gone?' (Vol. 53: 339) Graham Ashworth stated in 1967, that 'the wildman has his place as an inspirational source'. But it would seem that the new wild women, the feminists (the new brat pack? p.146), were invisible in their world view. Catanese's article of 1970, 'where is the planning profession heading?' (Vol. 56: 6) compared the British and American systems, but made no reference to the European community, or to women. In 1982, (Vol. 68: 184, September/October), Sorenson in an article entitled 'Planning comes of age' commented, 'planning's history is a story of unfulfilled ideas and utopias, of well meant but bitter failures'. Town planning had become quite powerful in the 1960s, particularly in respect of its associations with regional economic planning under the Labour administration of Harold Wilson. This was transient power from outside, reflected upon the profession, but not possessed internally by its members. The 1971 Town and Country Planning Act gave a new lease of life to town planning. Planners were encouraged to undertake surveys of the human activities within the urban system that created the land uses in the first place. But planners only had legislative powers to control the latter, not the former. Public participation became a more important part of the planning process, possibly as a sop to public outcry about top-down planning. But who did the planners see as being the public? A cartoon accompanying an article on public participation in 1973 (Vol. 59: 167) shows a bus full of men all wearing little bowler hats. Everybody, except the planners it would seem, knew that most bus passengers were women and that men with bowler hats used taxis! In the same issue, on page 171, the journal comments that Maureen Taylor and Peter Stringer did a random sample of 200 'adult women' (sic.) to find out what they thought about plans. In 1975 (Vol. 61: 99) an item informed readers that 'members of the public are given information about the alternative plans for Cheshire', which encapsulates the realities of public participation.

A MORE REALISTIC VIEW OF SOCIETY?

Race and space

From the mid 1960s the planning profession became troubled by, and often blamed for, urban social unrest. A concern with inner-city issues and racial problems was to open a path along which gender issues could also travel (see Appendix II, p.198 for texts). A series of studies of racial issues was undertaken, such as the work of Rex and Moore (1967). They made considerable use of women research assistants, As Moore (1977: 101) commented, he spent all his time 'putting cold, weeping girls back on the bus to Sparkbrook'. There is scarcely a gender element in the definition of the housing classes defined, and little differentiation of the problems of race in respect of gender. This white, male sociological perspective to racial issues manifested itself subsequently in the astoundingly womanless report *Planning for a Multi-Racial Britain* (RTPI, 1983; also see Vol. 70: 26, December 1984, 'The black dimension' by Aloke Biswas and Dick Simmons). But Rosa Parkes' decision to sit in the white part of her bus helped galvanize the equal rights movement in the USA (surely a zone-zapping action and a transportation issue). Is the new PTPI report (1993) any better?

Planners were more likely to see black people as exotic, or a colonial issue (Vol. 44: 137, May 1958 on Ghana; and Vol. 43, No. 3: 65, February 1957, on the then British West Indies). In March 1957 (Vol 43: 96) a book review on *Human Relations in Inter-Racial Housing* by Daniel Wilner and Rosabelle Price Walkley (1955) discussed the effects of the US Supreme Court's decision to require racial integration of public housing. This sought to reverse years of racial zoning. This issue is more commonly associated with South Africa, where apartheid sprang from the same North European roots (through Afrikaner Dutch colonization) as land-use zoning (Mabin, 1991). Holland had, for many centuries, sought to accommodate a mixture of religious groups within its limited land space, the civilized solution being to allocate different denominations separate areas to enable 'conflict avoidance' whilst living in close proximity – to permit each community to 'live and let live'. This was achieved by a policy known as *verzuiling* in Dutch (*zuil* meaning pillar (arguably aligned to *zuiver*, purity) Van Veen and Van der Sijs, 1991), that is 'pillarization' – the meticulous compartmentalization of people, objects and ideas. Pillarization is from the Latin *pila*, a word which was also used in West Germanic and Old French to denote a landmark, pillar, or pile of stones marking a boundary as an expression of order and ownership. Paradoxically the idea of separation spread widely, because of the mixing of ideas and words within different language groups. Both apartheid, with its emphasis on separate development (of races), and land-use zoning based on separate spheres (for each gender) may be seen as extreme outworkings of pillarization. The separation was not equal but associated with domination of one group over the other, with one side benefiting at the expense of the other, often with a large gap economically between the two sides. Planners such as Sitte (see p.115), Otto Wagner, and Joseph Stübben, author of *Handbuch der Architektur*, written in 1907 reflected such zoning ideas in their work (Collins and Craseman-Collins, 1965). The influence of these Germanic planners' ideas may be seen in the work of Geddes, Adshead (cf. 1923: chapter 8) and Abercrombie (1943).

Returning to the 1960s, in dealing with urban problems, British planners continued to put different aspects of life into distinct compartments (pillorization?), not making lateral connections and thus not realizing the likely racist (and sexist) implications of their land-use policies. In 1963 (Vol. 49,

No.8: 276) a CPO (compulsory purchase order) was declared on 60 acres in Brixton, South London. The CPO was announced in quite a matter-of-fact way, and did not mention it was an area where many Afro-Caribbeans were making their home. See-sawing from one reality to another: it took me a long time to realize that what were referred to as (those poor deprived) 'ethnic minorities' at college, included the self-same spiritually powerful people who had been such an inspiration to me in the black churches with which we had fellowship in the vicinity – who, in contrast, saw themselves as the sons and daughters of the living God. After the Brixton riots of the 1980s, the Scarman Report (1982) identified planning blight in the Coldharbour Road area as one of the factors which contributed to social unrest. Outsiders saw the British planning profession as insular and ignorant regarding racial issues. Suzanne Keller, writing her 'Impressions of St Andrews Summer School' (Vol. 51: 369, 1965), stated how parochial and cosily 'English' it all was. She was working with Doxiadis (1968), but later became a leading urban feminist (Keller, 1981). Nowadays, many see Europe itself becoming inward-looking and racially exclusionary with the enforcement of EC (European Castle) immigration policies.

Real working-class communities?

By the late 1960s the emphasis on slum clearance was being replaced by a concern with preserving communities and housing improvement, in no small part because of the efforts of residents groups fighting the planners. Nicholas Taylor wrote on the importance of 'The village in the city' (Vol. 59: 435, 1973). He was described as one of the new generation of Labour councillors who swept to power in the London borough elections of 1971 (Barnes, 1989), he was an architect and early gentrifier (Taylor, 1990). The January 1975 (Vol. 61) issue of the journal is devoted to urban conservation, with pictures of the Clifton conservation area in Bristol. Conservation was not a socially neutral process, because as areas went up in value, working-class residents found themselves pushed out. Ventures such as the SNAP project in Liverpool (1972) and GEAR in Glasgow (Taylor, 1986) received a mixed reception from residents. It is significant that some women who were later to become active in the urban feminist movement were involved in such schemes (Morris, 1986). A range of specialist area-based policy initiatives developed, concerned with revitalizing the inner city. These ranged from GIA policy (general improvement area) for upgrading working-class housing areas, through the conservation area of the 1970s, to the commercially inspired idea of the enterprise zone in the 1980s. Whilst such an approach enables resource targeting to such areas, it also picks them out as deficient.

Women's alternative perspective was beginning to be acknowledged in planning circles. Florence Rossetti's article on community care in the inner city in 1973 (Vol. 59: 361) identified specific problems about working mothers and pre-school childcare. She discussed the provision of women's jobs as a distinct issue from male employment. This article is feminist in all but name. The publication of *The Estate outside the Dwelling* (D.o.E, 1972c) (discussed in Vol. 59: 4, 1973) was a landmark. It is the first Department of the Environment Design Bulletin which investigated women's perceptions of the planning of residential areas (reiterating the ongoing insides/outsides debate). In it women were chiefly referred to as housewives, but at least the study provided a vehicle to draw attention to the problem. The traditional specialism of estate design was entering a new phase in which spatial standards were being given a new status, after the somewhat anti-spatial phase of the 1960s. A series of studies were made in Britain and North America of crime and vandalism on new housing estates. The emphasis on the need for 'defensible space' (Newman, 1973) detracted from women's design needs. The delineation of defensible space created new hazards, such as boundary walls which interrupted the field of vision and could be used for ambushing women. The sociological fraternity seemed to express more sympathy with the frustrations of violent adolescent boys than it did with women as victims. The physical nature of the area might be blamed for making people commit crime and vandalism, as reflected in the theory of environmental determinism which is so non-gender specific.

Studies questioning the actions of the urban professionals themselves, as against their subjects, were becoming more acceptable, such as the work of Pahl (1977), which showed that all was not fair or value free. Urban managerialism, including town planning, was seen as a form of social control. More radical studies of urban processes, and professional groups emerged (Harvey, 1975; Dunleavy, 1980; Simmie, 1974, 1981; Goldsmith, 1980; and Bassett and Short, 1980). Bailey's *Social Theory for Planning* (1975), was reviewed by Simmie (Vol. 61: 165, 1975). In this, Simmie raised the key question of how social theory might be put into practice, thus prefiguring the praxis debate. Earlier studies had highlighted the growing power of the professions as a social group within Britain (Millerson, 1964). Thompson's book (1968) on chartered surveyors had been reviewed by R.S. McConnell (Vol. 54: 452, 1968), who remarked on his over-preoccupation with the social status of the surveyors. I presume the urban professionals under discussion in such books were male, as women are seldom mentioned (Dunleavy, 1980: 115).

Rise of the new left

By the 1960s town planning was ripe for colonization by the new left, who saw it, and the untapped legislative powers with which it is imbued, as an ideal vehicle for changing the very structure of society. The move towards a more conflict-related view of society was informed by inner city unrest, and neo-Marxian sociology: the latter's influence extending beyond the realms of sociology and affecting geography and town planning (Pickvance, 1977; Castells, 1977). Marxist ideology fitted well with planning's emphasis on the worker and the public realm. The planners, previously Plato's guardians, were reincarnated as the political elite of Marxism, whose plans would eventually replace market forces. Such theory is replete with images of conflict and struggle which put the planners as heroes and saviours at the centre of an epic, urban battle. Apparently, nothing could be done until 'after the revolution', absolving the planners from the need to act on women's 'trivial' urban demands, or racial problems, in the now. Various women were realizing what was happening; in 1974, Patsy Healey, then a senior lecturer at Kingston Polytechnic (and now Professor of Planning at Newcastle University), in an article on 'the problem of ideology' (Vol. 60: 602) 'names' the shift away from utopianism to Marxism. The emphasis on impersonal economic forces, combined with references to an abstract working-class, meant there was little place for real people, let alone women or geographical space, in the theoretical debates (echoing ancient spirit/body dichotomies). Castells' impersonal view of the city and the individual as nothing more than a unit of labour power was alarming (1977). (Subsequently he wisely revised his views, and stressed the human dimensions of consumption (Saunders, 1985).) One was made to feel very small, and bourgeois, if one mentioned women in this setting. The place of women was somewhat ambiguous, as they were neither land nor society. Sometimes one felt they were property, as the suburban housewife seemed to be plumbed into the house along with the washing machine in much neo-Marxian theory on housing classes (as in Bassett and Short, 1980).

Later neo-Marxist theoretical developments allowed emphasis to be put on consumption, albeit defined in male terms. Saunders (1985: 85) adopted a more rounded sociological viewpoint in which urban issues were looked at from the perspective of the urban resident actually living in the area, who consumes goods and services such as housing, infrastructural services, schools and shopping provision within the community. Previously, emphasis had been upon the world of the worker and capitalist involved in production in the workplace away from the home. The emphasis on consumption opened the way for women to redirect attention to the urban political significance of the domestic realm and the residential area as a valid place for serious class struggle, and to redefine production and consumption from a feminist perspective. Many disadvantaged urban groups started thinking and working for themselves to press for the sort of cities and society they wanted, rather than accepting the views of academics and professionals as to how they should live. But it takes a long time for official academic theory to catch up with them, and 'appropriate' their ideas. Planning had to run through a phase in which class elbowed in and took central stage, before gender was eventually allowed space. It seems to me the second wave of feminism was well under way before neo-Marxism grabbed the limelight. Any patriarchal theory might have served the same purpose to detract from women's demands. Freudianism had a similar role in detracting from first-wave feminism, making the agenda one of sex not gender: one of individual, psychologicial health, rather than of cultural, societal change. Many community groups felt Marxism could not help them; they saw it, like traditional town planning, as an elitist top-down middle-class conspiracy. There was often a contrast between the importance planners put upon class as an abstract issue, and the obvious embarrassment and gaucheness they manifested when talking to members of the working-classes – face to face.

Ethnicity, and disability too, appeared to be stronger initially than gender as factors which legitimated the need for the development of special policies for groups who are different (Greed, 1991). By the mid 1980s it became acceptable to stress the social needs of special groups rather than grand theory for society. Disability was highlighted as an issue by the GLC, which had now taken on a new left persona. Although this was a welcome initiative, it was a potentially two-edged sword, reflecting dichotomous thinking based around perceived differences between normal/special, abled/disabled, and vast majority/tiny minority needs. Disability policy can be conceived in planning circles as being solely to do with difficulties getting up steps – so all that is needed is a few ramps. An article in January 1984 (Vol. 70: 5) discusses 'planning for the handicapped', and is more positive compared with earlier incidences which tended to centre around dropped kerbs and men with battery cars (Vol. 68: 45 et seq., 1982). The Town and Country Planning Association presented a more enlightened approach, in a special issue of *Community Network* (Vol. 8, No.3, Autumn, 1991) entitled, 'Access for All'. The first substantial article on the disabled in *The Planner* appeared on 7 August 1989 (11: 10–11, weekly issue) describing the Manchester Disability Study by Phil Barton and Ian Thwaite. There is a danger in treating 'special groups' as separate from mainstream society, when often their plight is, in part, caused by the deeper economic and social structure of society.

Neo-Marxism eventually went out of fashion, and the whole paradigm shifted again in the subsequent post-Marxian, post-modernist phase. Weber's ideas came back into favour in some quarters – a mixed blessing for women (Bologh, 1990). Although women planners might be treated pleasantly (smoothly) their requests for the implementation of equal opportunities policy are still not taken seriously. Within the enterprise culture of the 1980s, the property professions, including some planners, became more concerned with market forces, which left little space for the consideration of non-profit-making gender issues. Many women are now more concerned about dealing with market-led 'managerialism' which appears to be as sexist in mission and ethos as Marxism ever was. In recent years there has been a retreat from the emphasis on heavy deterministic theories, and a greater acknowl-edgment of the complex variety of factors, over and above economics and class, which can influence people's life experiences (cf. Hall and Jacques, 1989; Hamnett et al., 1989). There is greater awareness of the effects of personal characteristics (such as race, gender and home locality) on a person's position in, and experience of, society. But, the new relativism of post-modernism can make gender just one interesting factor among many (McDowell, 1992).

One of the problems with neo-Marxism was that although it could provide a theoretical framework, it was difficult to translate this into concrete policy. This was particularly a problem for planners. The overriding emphasis on economic and political forces made any physical intervention in the built environment itself appear as mere superficial tinkering with the super-structure. With time, neo-Marxism was virtually forgotten, new 'brat packs' and theories emerged, and the pendulum swung away from aspatial planning and back to spatial planning. A new enthusiasm was manifested for urban design, this being linked to the movement for conservation of historic areas. Whilst planners in the 1960s looked to the future, and in the 1970s to social revolution, by the 1980s there appeared to be a desire to recreate the past as reflected in post-modernist, and neo-classical architecture. The urban design movement may be seen as redefining the discourse of planning as art. It manifests a greater preoccupation with the physical aspects of the built environment than with the underlying social and economic factors which generate the built environment in the first place. But the emphasis on aesthetic design may marginalize women's issues (cf. p.65–9). The return to traditional values was linked to the rise of the enterprise culture, requiring a more market-orientated mentality from town planners. In the 1990s the green movement continues to grow in popularity, and the topics of sustainability and environ-mental assessment have taken the limelight (Fortlage, 1990). We shall yet see whether any of this is really any good for women.

Changing attitudes?

I would like to balance the above discussion of the effects of public-realm theory on women's lot with a consideration of planners' private views of women as betrayed by items in the journal. It seems to me that whilst planning theory might stress the importance of the working classes, and even women, the personal attitudes of some planners towards women, the working class, and indeed the general public, remained immature, and school-boyish. In 1972, the famed Grotton Saga began (Vol. 58: 276). This was a spoof series about an inefficient planning department (culminating in 'the book', Ankers et al., 1979). It teeters on the edge of both racism and sexism, and yet shows some disrespect for the stupider aspects of the planning system. The 'Dan the Plan' cartoon appeared in the mid 1970s, followed by the 'Polyanna' series. Such cartoons combined a mixture of sometimes nasty, student rag 'humour' with occasionally amusing social comment and insight. More disturbing was a strange series, 'Planners' Realm', (Vol. 62: 104) in 1976. This was a romantic spoof about 'love on site', all about Abigail the maggot breeder's daughter, and 'the mysterious Sylvia from the past'. Sylvia is the name of the only woman president of the Institute. Much of this was written as if it was assumed the only readers of the journal were men, a spoof advertisement for a free tobacco pouch appearing with this series. Other little things still give a male feel to the journal, such as the use each year of photographs illustrating the annual conference which usually only show crowds of male delegates. Such images can undo my efforts to convince female students that really planning is not such a predominantly male profession as they imagined it to be.

Coinciding with an increase in the numbers of women entering planning in the 1980s, one does get more photographs of women in the journal but not in a positive manner. In a series on the need for the local authority planner to present a more professional image (*Planner News* throughout 1983), one issue (December 1983: 4) shows a receptionist, located on the page alongside a photograph of Audrey Lees, who was virtually the only woman chief planning officer at that time: the contrast in types is stark. The culmulative effect of endless little instances, a general ethos of ignorance about women, and constant allusions to football and drinking (Hey, 1986) is wearing. Individual women have attested to office experiences ranging from mild innuendo and jokiness to extreme sexual harrassment. Women planning students report the continuance of negative comments about housewives and women by some lecturers. The situation is meant to be better nowadays, but only recently women students at conferences have described to me the frequency of such problems in modern schools of architecture and planning. All is not well in the private realm of womens' personal experience, in spite of the improved public image.

10

WOMEN INTO PLANNING
Ways and means

THE OPPORTUNITY OF POST-WAR PLANNING EDUCATION

Expansion and credentialization

In this chapter I discuss the factors which facilitated the development of the current women and planning movement. In the first section I give a short overview of the nature of planning education since the Second World War because it provided opportunities for women, and a means for other ideas, to enter the profession. In the next section I discuss the types of women who are the most effective as zone zappers to implement women and planning policies. But the circumstances for change had to be right too. It would seem that the raising of 'women and planning' issues became more acceptable within the context of the rise of new left politics, particularly within certain Labour-led local authorities in the 1980s. In the last section I describe the development of the current women and planning movement, as manifested in policies and initiatives adopted by planning offices, my intention being to give an overview, not a detailed account, using selective examples. As stated in chapter 1, I consider it is very important to include the names of wonen planners to redress the balance of the historical record which has mainly excluded women. Town planning is a classic example of a specialist profession that developed as a composite discipline, drawing elements from surveying, architecture, civil engineering, public health and housing. At first, no one was admitted unless they were prior members of such professions (p.108). As planning became established as a separate profession, to formalize its status it went through a process of credentialization, which is the process of limiting access to specialized professional enclaves by increasing the educational requirements (Collins, 1979: 90–1).

In 1946 the first undergraduate degree course was established at Newcastle (Vol. 32, No.6: 224, and see my footnote 4 p.17), based on five years' full-time; quite a leap from a few hours a week for a year or so! The aim was to make the length comparable to architectural courses to give similar status. The course

156

welcomed war veterans (p.132), who often got exemptions as to the length of course, and 'boys and girls from school'. The subjects included economics, sociology, law and geography, and constituted an attempt at creating a new type of purpose-built planning course. Subjects such as property valuation, geology, mining, engineering theory and public utilities crowded the syllabus, as if they were allowing for every eventuality. Making mathematics a first-year subject was a good way of 'weeding out the girls'. This might have been the result of influences from North America, where it was customary to teach calculus to civil engineering candidates, including planning students who were unlikely to need it. The emerging important planning colleges in the States included MIT, and other high-status Ivy League colleges. Princeton began a rather refined version of town planning alongside architecture. Princess Noor of Jordan studied town planning at Princeton; it appears to be a royal preoccupation (Prince of Wales, 1989). In North America more school leavers, including women, were going to college, and this enabled more women to enter town planning, some such women producing foundational texts for the movement (Appendix II, p.198).

In contrast, in Britain, only a small minority of people took undergraduate professional degrees in the immediate post-war period. The pupillage system of articles was still strong in the professions. Many young grammar-school boys went straight into local authority work at 16 to train to become planners, as advertisements in the journal attest. The RICS surveyors, opportunistic as ever, were also offering a town planning examination option, with part-time and correspondence courses overseen by their College of Estate Management. There appear to have been very few women on early post-war undergraduate courses, because of the double hurdle of having to be previously qualified in another profession such as architecture, and it is not until the 1960s that women enter planning courses in any numbers. As with surveying I would describe the growth in the number of women as a 'splutter effect' like a fire sparking up, and then dying down several times before it gradually draws up and roars away, (Greed, 1991: 91). From college records, and reminiscences of women planners, the pattern in planning appears to be more 'extreme' than in surveying. In the 1960s one might get no women on the first year, then perhaps 20 per cent the next year because someone miscalculated the intake figures and panicked, then no-one again for three years, then two or three coming in, and maybe one of those leaving out of exasperation. Taking a typical planning course this might work out to say, 3–5 per cent of the total student body, but with considerable variation between years. After this initial, unstable stage, there might then be a slow, gradual, build up culminating in a 'catching fire' in the late 1970s and 1980s, when numbers shoot up to 30 per cent or more. I suspect in the 1960s there was an element of discriminatory selection against women when dealing with applications and that there was really a greater demand amongst women than is evidenced by the intake figures.

With the introduction of a comprehensive town planning system in the 1940s the government had been caught on the hop without enough planners to run it (Eve, 1948). Discussions of planning education were apparently couched in non-sexist phraseology, career information frequently referring to 'a boy (or girl)'. Likewise Gibbon (1942: 199) comments, 'the idea apparently is that the central authority should train a pool of planners. From this pool men (including women, as throughout) would presumably be drawn for the work of the central department'. Consequently the Schuster Report (1950) on planning education (Vol. 36, No.8: 281, 1950) made recommendations on how the courses should be organized. Schuster states the need for 'certain innate qualities of intellect and character', and discusses how these qualities might be produced. Three requirements are stipulated: (i) a sound basic education; (ii) some specialist education for planning (only 'some' note); and (iii) practical training and experience. From a feminist perspective these requirements are as worrying as the surveyors' desire to recruit 'the right type', who also is obviously a good chap – one of the property fraternity (Greed, 1991: 74). Meanwhile the professional body was gaining status, receiving its Royal Charter in 1959. It was normal for most would-be planners, like surveyors, to study by part-time routes (Henniker-Heaton, 1964). The expansion of higher education in the 1960s enabled more women to enter planning on merit rather than on the basis of town hall patronage. The policy was intended to increase opportunities for working-class males, although both middle-class men and women were to benefit (Robbins, 1963; DES, 1966; Robinson, 1968; Lane, 1975). An item (Vol. 55: 24, 1969) on student opinions on employment opportunities stated that some local authorities were not keen on employing students who apparently were only good for 'mending maps and filching fags', seeing the new courses as a challenge to their own in-house training. Apparently, students felt that planners were poorly paid, stating, 'for girls where job opportunities are fewer, planning jobs are more remunerative than others'. Women were seen as taking men's jobs and reducing salaries, although they constituted around 1 per cent of the profession and less than 5 per cent of students. In fact job opportunities in planning were good and well paid in the 1960s because of the expansion under Labour of regional and urban planning activities. A woman, Miss Elizabeth White, was secretary of the Student Planners Association making these comments, a classic helper role; perhaps no one asked her views.

Ted Kitchen (1970) in an article, 'Planning education in Britain, the state of the game', comments that by 1965, thirty planning schools were operational resulting in a 75 per cent increase in courses in the 1960s, but a greater than 200 per cent increase in actual student numbers at undergraduate level, and a 100 per cent increase at postgraduate level. The number of part-time and direct entry courses went down to 25 per cent of all courses. He refers to research by Cynthia Cockburn on planning education (who was to become a famous urban feminist: Cockburn, 1983, 1985). More solid figures are given (Vol. 57: 54, 1971) by Susanna Marcus (1971) in her famous article 'Planners – Who are you?'. Marcus found that most planners were middle-class, and that prior professionals (p.108), who then constituted 45 per cent of planners were more likely to have professional fathers than graduate planners were. There were few women unearthed in the survey and even the work of women researchers in those days appears non-gender specific. *The Planner* frequently returned to the theme of non-participators and non-joiners (for example, Vol. 60: 559, 1974) expressing fears about the future shortage of planners. It would seem many older planners had simply fallen into planning as a good thing to do, without the strong ideological identification which was a feature of the new generation. In contrast, I remember how difficult it was for me to find out about planning courses. When I did become a planning student I was informed I was 'not interested' in the course (when I was very keen), suggesting I gave out totally the wrong subcultural signals. See-sawing: resentfully, for my part, I found it very hard to believe that in order to solve the problems of the inner urban poor it was necessary to create and pay an expensive professional class who were so ignorant of people's lives that they actually had to go to college to learn about the urban situation before they could even start solving our problems for us!

Subsequent studies of the composition of the profession were undertaken by Linda Walsh and Mike Gibson (1985) and recently by Vincent Nadin and Sally Jones in 1990, at which time it was found that only 1.4 per cent of planners were taking the direct-entrance non-college route. Women researchers have always been in the forefront of questioning planning education and the belief system of the profession, and may be seen as a form of philospher queen, vital to the development of the women and planning movement. Josephine Reynolds asked, 'are there too many planners?' at a controversial conference on planning education also attended by Sylvia Law and Cynthia Cockburn (an interesting combination) (Vol. 57: 132, 1971). A Miss Minett asked, 'Is planning a profession?' (Vol. 57: 231, 1971). Women appeared freer, as outsiders, to question whether it was, both Minett and Cockburn being quoted (albeit ungendered) in discussions on the future of the profession (RTPI, 1971: 31).

Proliferation and demarcation

Nowadays planning courses are provided at many levels, and hundreds of planning students are convinced that it *is* a profession, although their reality is based on a set of flimsy beliefs, which have been convincingly passed on to them as truth by lecturers and members of the profession (cf. Clark, 1983). Planning courses are still hedging their bets as to content, incorporating a wide range of topics: so much so that the solution to fitting it all in is modularization, in which the students elect for particular options. Undergraduate courses nowadays are likely to include social and economic analysis, history and theory, environmental design, development control, development plan, development process, and various specialisms such as rural planning, transportation, local planning, planning methodology and management studies. Add to this a language and a European dimension in some cases. Around a third of planning courses are located in polytechnics (since 1992 renamed new universities); they have constituted the biggest growth area in terms of student numbers, are often more progressive in attitude, and have some of the highest percentages of women students, up to 40 per cent, whereas the average is nearer 30 per cent or less. There are around fifteen undergraduate planning courses, of which five are in polytechnics with a total first-year intake of around 526 per year, whereas there are twenty-one postgraduate courses, mainly in universities, with an intake of around 700 per year (Vol. 78, No.8: 19–20, 17 April 1992). Under 10 per cent of academic staff are women, but some courses have fewer than this, and one professor proudly replied to my enquiry by stating that his department had three female secretaries! Many planning courses consist of an initial three-year undergraduate degree, followed by a year out in practice, and then another year in to give full exemption from the final examinations of the RTPI. The situation is further complicated in terms of full-time and part-time options, modularization, and a variety of qualifying short courses. On some full-time courses nowadays, many students are really part-time because they have outside jobs. One should not underestimate the power of the non-degree technician level of planning education, which still provides a possible vertical access route for those who cannot attend college full-time (Percy, 1970). Technician courses tend to attract a wider range of types of women, some of whom may prove to be inspired philosopher queens.

It could be argued that devaluation of the professional status occurred as a result of educational inflation (Dore, 1976). The increase in the numbers of women was seen as one of the causes of its tilting downwards (cf. Reed, 1990). The expansion was balanced by greater unofficial demarcation of different levels within the profession on the basis of academic, professional and personal criteria. Whereas for years planning was overwhelmingly male, it might nowadays be seen, relative to other construction professions, as more female, as a compensatory social land use profession. But, the academic nature of planning might be seen as more rigorous than surveying. There are very few post-graduate surveying courses. Some would say that in the future town planning should only be a postgraduate profession (as it was originally), or that progression to full professional status should be limited to those with upper-second degrees or above (Vol. 78, No.10: 9, 15 May 1992), as is virtually the case in the legal profession. This does not encourage more open access for disadvantaged groups.

Limitations and opportunities

Whilst the expansion and diversification of planning education does give women a chance to get into the world of planning, as with women surveyors (Greed, 1991: 103) it does not mean they like their experience of planning education. They might have quite a different experience of planning education from male students, and from each other. There are considerable differences in the experience of elite women undertaking postgraduate planning degrees at internationally recognized schools, and those entering polytechnic courses at undergraduate level. After examining a range of college returns, I must conclude that, nowadays, there appears to be little overt discrimination in terms of student grades, but that does not mean women necessarily like the course, and does not discount the fact that they may be expected to work twice as hard to achieve the same mark, as they are not given the benefit of the doubt. Likewise, women lecturers have pointed out that even when equal opportunities provision is pursued on behalf of women students, for example crèche facilities, women lecturers are not necessarily also eligible. The purpose of several EO policies appears to be to increase student numbers – that is, non-gendered raw material – not to increase or retain the number of women lecturers, who may be seen as being in competition with the men. Planning educationalists are concerned with demand and supply (Dickens and Fidler, 1990). For texts on women's experience of built environment education see Appendix II. p.200.

Women seem to be better qualified than men. According to the RTPI report *Planning for Choice and Opportunity* (1989b: 101), in 1987, 42 per cent of women gained first-class or upper-second degrees as against 30 per cent of men, but there appears to be little correlation between educational achievement and progression, because of the roles expected of women. The question of whether postgraduate planning students do better career-wise has been the subject of much debate (see letter from Bernard Evans, Vol. 77, No.20: 4, 7 June 1991 'Undergraduate v. Postgraduate'). Seventy-seven per cent of planners have an undergraduate degree, a third of which are in town planning, but this proportion increases towards the younger age groups, the second most common degree being geography (Nadin and Jones, 1990). Only 19 per cent of planners are qualified to master's level or above, with around another 5 per cent having research degrees and other higher qualifications. In contrast, only 1.4 per cent nowadays qualify through direct entry (compared with 20 per cent twenty years ago, Marcus, 1971). Women have more degrees, for example 57 per cent of women planners in Scotland had two or more degrees (Brand 1985). In Scotland, 9 per cent of women working in planning offices appear to be in more senior grades than in England (national and regional cultural differences within Britain cannot be underestimated). It would seem that highly motivated academic women can form a powerful elite and succeed more in Scotland than England, relatively speaking (cf. Delamont, 1976;

1989). It is no coincidence that some of the main leaders of the women and planning movement are Scottish, for instance Hilary Howatt and Janet Brand and the chair of the women's panel in 1992 was Una Somerville from Ireland. Women who took planning at a postgraduate level have been among the leaders of the women and planning movement and are more likely to be found in policy making areas rather than development control. Women with a first degree in geography have provided much of the initiative for the women and planning movement possibly because they were exposed to a wider, more humanities-based discourse before being subjected to professional socialization to make them planners. But they are not necessarily more senior. Some women may end up on the lower levels in this segmentation process, irrespective of their academic abilities, as a reserve labour force. In spite of employers producing equal opportunities statements, women planners are not getting as far as they should because the criteria on which EO is based are often highly gendered (Collinson, et al., 1990; Halford, 1987). The literature on such problems is extensive for a range of other professions (see Appendix II, p.200). Those with the wrong personal views or attributes are unlikely to progress within the subculture and reach senior decision-making levels where they can influence the nature of the built environment through women-centred policy making. Nevertheless the situation in education is generally seen as being more liberal than that in practice, where women come up against old men who still may never have heard of equal opportunities. The RTPI itself now requires colleges to give particular attention to giving students and appreciation of EO issues and diversity (RTPI, 1992)

ZONE ZAPPING: THE WOMEN AND PLANNING MOVEMENT

Likely types

Women within the built environment professions are not a unitary group (Greed, 1991: 10–11) and it would seem that each type has a contribution to make in changing town planning. As indicated earlier, those women planners most likely to succeed as zone zappers (but not necessarily in personal career terms) are the philosopher queen and the femocrat, but others including the bourgeois feminist have an important role to play. The women and planning movement is also dependent on the support and voice of many ordinary women, who are not planners but who have been involved in various community groups in fighting the planners (Aldous, 1972); or who are concerned with design issues related to their role as carers (Leach, 1979). The adherents of the first-wave of urban feminism included women visionaries and dreamers (chapter 6), many of whom were writers, who may be seen as philosopher queens. Second-wave philosopher queens were more likely to be found working in academia, and in particular women with a background in geography or sociology had a valuable role in conceptualizing women and planning theories and policies. Women in society as a whole were being influenced by the new wave of feminist books coming from North America and Australia (from such philosopher queens as Friedan, 1963; Millett, 1970; Greer, 1973). Women planners seem to cultivate wider intellectual and academic contacts and networks than men, often coming from humanities and arts backgrounds and knowing people from a range of academic disciplines, so they bring with them to town planning a greater mix of alternative views. Modern urban feminist literature is outlined in Appendix II. Women acting as researchers (typically as assistant to the male researcher) have also been able to have an influence on the knowledge base of the profession. A few other women have acted as philosopher queens outside academia, for example the American Jane Jacobs (1964), who started as a journalist not a planner.

Whilst philosopher queens might have developed visions and ideas, the next stage was to change the urban situation. It seems to me that those who are achieving the most seldom write their ideas down, because they do not have time or because they see themselves as practical rather than academic. Elaine Showalter (1982: 11) has described women as sociological chameleons taking on the class, life style, and culture of their male relatives and associates to survive. Paradoxically, an essential attribute for members of minorities is to be able to blend in, to pass, and not appear to be different. It would seem that many of those who want to succeed as zone zappers give a convincing appearance of being suitable to fit in to male structures, and apparently possess the right so-called managerial qualities to work in the (male) team. For the purposes of getting into local government, the femocrat (woman bureaucrat) seems to be one of the most enduring and acceptable types of woman planner (Leoff, 1987: 14; Watson, 1990). (Incidentally, Sophie Watson subsequently became professor of town planning at Sydney University.) In the selection process the haloes/horns effect seems to operate especially for women; they may be seen as all good or all bad (*vide* virgins/whores), leading to the stars/victims dichotomy. A few acceptable successful women surge ahead, whilst the rest are encouraged to blame themselves, not patriarchal harassing sexual power-structures for their failure (cf. Merchant, 1993). Whilst the 'good girl' femocrat might be effective in getting employment, there is a need for more radical 'bad girl' urban feminists to get things going in the first place. These are often found on the fringes of state planning, possibly involved with community groups. Such a woman may achieve policy breakthroughs to the benefit of all women but, once used, is often passed over at the first opportunity in favour of her moderate sisters – especially by careerist male planners concerned with their image.

Whilst expression of a certain amount of municipal socialism and liberal feminism, or domestic feminism (for other women), might be expected from the femocrat, women who flout other more extreme political, religious, or sexual ideologies might be seen as too unsettling or not suitable to work in teams (a frequently reported sexist criticism). When neo-Marxism was fashionable in the 1970s and early 1980s it would seem that one of the types of woman most likely to succeed was the woman academic who appeared to embrace the neo-Marxist cause. Planners may be expected to be anti-establishment, and radically socialist but only within certain (male) parameters. Within the context of municipal socialism women were more likely to be welcomed if they could show they were efficent managers than if they were revolutionaries, visionaries or academics. Some of the most successful women were not feminists. Even if they are retrospectively associated with feminist initiatives, it might not have been supportive of other women at a personal level at the time. As Levine (in conversation, and cf. 1990) commented, some women who lack power over men use their individualist, careerist feminism to exercise power over other women. However, one also hears stories of femocrat planners who, by advising and encouraging junior women in their employ, help them up, some to subsequently star as senior women themselves. But as Mies (1987) points out, women planners are not necessarily any more pro-women, or more socially aware, than men. The converse (albeit rare) is the male feminist who seeks to promote women's interests. One cannot judge too much because individual women often have great difficulty knowing what is expected of them, and reconciling public/private dichotomies in their own lives, and 'getting by' (Riley and Bailey, 1983). If something is right for a man, when a woman does the same thing it might be taken quite differently. Women planners experience two sets of socialization within our culture (Hite, 1988: 132; Sharpe, 1976: 176): to be women and later to be planners. Professional socialization is arguably an extreme form of male socialization, bringing out the worst aspects of competition and aggression in women towards others (including other women).

I found in my previous research that one of the most popular types in surveying was the bourgeois feminist, the businesswoman who wanted the independence of a career (Greed, 1991: 22; Hertz, 1986), within the context of the enterprise culture. I consider that as such the bourgeois feminist can act as the zone zapper *par excellence* in promoting women's viewpoint on built environment matters in the commercial property professions (Greed, 1991), but she has a limited though still important place in the planning profession, in which less than 20 per cent of members are working in the private sector. During the property boom of the 1980s, some young women planners eschewed local authority employment altogether for the private sector, as Professor Janice Morphet, one of the few women to be a town planning professor and the chief planner of a local authority (sequentially), pointed out at the 1989 RTPI Annual Summer School Conference (Vol. 76, No.7: 58). Even in the recession-ridden 1990s, those women who were teenagers within the enterprise culture of the 1980s are generally more entrepreneurial (and less bureaucratic) in perspective, and so are less inclined towards public sector employment, particularly when they perceive a lack of real equal opportunities therein. This may lead to fewer women being found in local government in the future who can follow on from the pioneers of the second wave of feminism. These contrast with the more laid back, socially concerned type of women town planners who entered planning in the 1960s; such women have told me they are terrified by the new generation, 'in their business suits'. The bourgeois feminist may be seen as a descendent of the trader (Boulding, 1992: I, 227, and see pp.81–3), often an outsider, and it would seem that some women find that being self-employed (for example as a planning consultant) gives them more freedom and flexibility than they would find in public employment. Miller and Swanson's work (albeit 1958; see Greed, 1991: 32) crystallizes the dichotomy between bureaucratic/entrepreneurial types among professional women. The philosopher queen might take the role of entrepreneur, as did Henrietta Barnett, in order to put her ideas into practice. Such a role might be an option for modern women planners, but would be more likely to be feasible in America than Britain.

How it came about: penetrating the profession

For many years there appears to be little mention of women's issues in *The Planner*, or in local planning authorities, although there is, in fact, an imperceptible build-up, with such topics, initially, being referred to by code, variously, as social aspects, human needs, community politics or neighbourhood planning. In 1970 (Vol. 56: 399) Margaret Willis had written on the social content of structure plans whilst a senior research officer at the Ministry of Housing and Local Government (also contributing to D.o.E, 1972c). The 1970s constitute a watershed between the old and the new. In 1974 (Vol. 60: 918) the journal commented, 'critics like to perpetuate the image of the Institute as the rather faded lady of our crest. This is nonsense we are a modern forward looking institute, responding to change.' (And so?) in 1974 (Vol. 60: 914) Sylvia Law became the first woman president of the Institute, at which time she stated, 'I certainly do not intentionally pursue a feminine approach to planning', although she subsequently stressed the importance of human needs (and replied graciously by letter to my research enquiries). A prophetic photograph (Vol. 60: 967, 1974) shows her standing with the new GLC Chairman, David Pitt, the first black person to hold the post, but there was little mention of either ethnic or women's issues in the journals for years to come. Meanwhile Cynthia Cockburn produced a key book on the importance of community politics (1977). She had previously worked as a planning researcher, in association with Ted Kitchen and Sylvia Law (an interesting mixture). In 1976 (Vol. 62: 129) Audrey Lees became the first woman county planning officer at Merseyside, and subsequently Controller of Transportation and Development at the GLC in the early 1980s. Christine Hoyle, a graduate of Durham, became the first woman chair (anywhere) in Yorkshire, but died at the age of 41. Other percursors of future trends appeared such as in March 1978 when Honor Chapman (a senior surveyor) spoke to the RTPI about business (a dirty word in the planning subculture), possibly prefiguring the rise of the bourgeois feminist.

By the early 1980s, the women and planning movement was emerging more visibly.Carol James, a planning lecturer (who died in 1990), was one of the key figures in this rebirth, undertaking some of the first research on women and planning in the early 1980s at the Polytechnic of Central London (PCL, now University of Westminster) along with her colleague Jane Foulsham. Beverley Taylor, their research assistant, was a founder of WEB (*Women and the Built Environment Bulletin*), whilst Jos Boys and others who were to become key figures in the movement were also linked to PCL. A seminal women and planning conference was organized at PCL in 1982, which many saw as a key point in the development of the movement (see cover photographs of Vol. 70, No. 3). One of the first articles on the subject appeared in the mainstream press *Planning*, No.479: 8–9, 30 July 1982. In passing: I was stunned to find that the *same* photograph of a group of three women planners used in this article was used in an article ten years later in the same journal, illustrating the need for training of non-professional staff in planning offices (*Planning*, No.971: 6, 5 June 1992). Meanwhile, in the United States in 1979 the women and planning division of the American Planning Association held its first meeting, and soon became one of the largest divisions (cf. Leavitt, 1981). The RTPI, in contrast, were not leaders in the field but there were certain groups which were more receptive within the purview of the Institute (in addition to women), including the RIG (Radical Institute Group). But such allies had to be utilized with caution because of the maleness of much radical urban sociology and new-left politics. Their interest in womens issues was seen by many women planners as merely a transient power ploy. By 13 January 1983 a working party was constituted (and given £750 for expenses) through a motion by Anne Goring and Robin Thompson. This eventually resulted in the 1989 document *Planning for Choice and Opportunity* (RTPI, 1989b).

In 1981 (Vol. 67: 100) Jackie Underwood (then of the School of Advanced Urban Studies, Bristol) wrote on development control, and in the summer of 1982 she was guest editor for the journal. Women journalists were increasingly to be found editing, and writing articles for the property press (Greed, 1991), and in the case of *The Planner*, Ann Satchell and Diana De Deney took editorial roles in the 1980s. In the wider media world Judy Hillman had established herself as planning correspondent for the *Guardian* and for the *Evening Standard*, Mira Bar Hillel becoming the property correspondent for the latter. Gloria Hooper frequently appeared in the mid 1980s as correspondent on the EEC for planning matters, being a MEP herself. Women organizers such as Alison Ruff, and executives such as Margaret Catran, and female administrators at the Institute have all played their part. Other significant women were the small number of mathematical women in specialist planning areas such as computers, and predictive modelling, for example Margaret Roberts (1974), and Susan Barrett. Roberts was the first woman to head a planning school, at PCL from the late 1970s to 1989. Hazel McKay originally has a mathematical background, and therefore (on her own admission) thinks linearly and logically in a way which men understand. Other women were beavering away in more traditional areas like housing, such as Jane Brooke who was part of the GEAR inner-city revitalization team (Vol. 67: 83, June 1982; Taylor, 1986). Others were concerned with the disabled, such as Anne Davies and Sarah Leighton Lockton, (Vol. 68: 55, 1982); this article was the first sighting of the dropped kerb phenomenon. Nowadays disabled planners are demanding integration not segregation (*Community Network*, Special Issue, Autumn 1991, Town and Country Planning Association). Some men were incorporating women's issues into their mainstream writing. John Newby (December 1985: 27) writing on local area plans, actually stated that more nursery provision is necessary. Although nowadays employment statistics on women as well as men are collated in the planmaking process, policy making still seems to be centred around male employment. The importance of subscribing to equal opportunities became official, as evidenced in the 1988 presidential statement entitled 'Closing the Gap' in which point 5 was specifically concerned with women (Vol. 74: 10). Hilary Howatt had previously written on women in planning (Vol. 73: 11, August 1987) in relation to the recommendations of the working party. There was an amusing cartoon of a little boy in a planning office (22 June 1989: 5) declaring, 'My mum told me to come along and help with some development control', reflecting the reality of the problems of school holidays. An obituary for Janet Tassell, Oxfordshire Assistant Chief Planning Officer (31 May 1991: 9) bore tribute to the career of another forerunner, who, incidentally, had studied geography at University College London and postgraduate planning at Manchester. Although 'women and planning' had become more 'overt', women planners continue with their 'traditional' roles, Pat Castledine undertaking planning aid (2 June 1989: 24), and Diana Fitzsimmonds (13 April 1990: 29–30) writing on planning education from a women's perspective; but the male predominance remains.

Overlaps between the professional and the personal

See-sawing back into the personal realm, it is inevitable that some women planners are married to, in relationships with, or otherwise related to other planners (cf. Brandon, 1991), so the public and private realms overlap in their personal lives. This may give women the advantage of being 'accepted' within the subculture, but at the same time they may be overshadowed by their husbands and fathers. For example, in 1971 Mrs. Geraldine Amos (Vol. 57: 191) gave a talk to the Yorkshire branch on 'does physical planning solve social problems?', which was prophetic for its time, but arguably she was overshadowed by her famous planner husband Jim. Twenty years later (RTPI,

SW Branch, 1991) the same Mrs Amos is actively involved in promoting the needs of the disabled in planning, having been confined to a wheelchair herself, travelling all over the country, with difficulty, by train. John and Jane Darke have both achieved a great deal for the cause of women and planning. Dahlia Lichfield is carrying on the family tradition of Nathaniel Lichfield. Also, the daughter of the architect who designed Brasilia designed an Olympic City for Brazil's 1993 bid for the Games. Some younger feminists, who appear to take for granted the fruits of twenty years of feminist struggle, are against what they see as 'nepotism' arguing that planner-wives should not work in the same local authority, or college, as their husbands. Bearing in mind the limited numbers of local openings for women, without leaving home, and the fact that such a rule suggests some sort of conspiracy or power on the part of the couple (or husband), this attitude suggests an abstract form of feminism, divorced from daily inter-personal realities and options. Greater negative power is more likely to come from cliques of patriarchal men who promote and favour each other, than from husbands who might be seen as outcasts by other men for marrying overt feminists. It is usually the wives who are expected to leave (Stone, 1983). The fact that a woman may have struggled hard to become a planner in her own right long before she met her partner, or may since done research in her own right, and that the two may disagree totally on every aspect of planning policy is not considered in such debates. But the woman is often left out when the glory is given, as in the case of women architects. Half of women architects have partners or husbands who are also architects (WAC, 1993: 5). Elizabeth Plater Zyberg is the other half of Andreas Duony, who jointly built Seaside (Mohney and Easterbury 1991) (Prince Charles's ideal plotlands town); Denise Scott Brown contributed equally with Robert Venturi to the design of the Sainsbury Wing of the National Gallery (Lorenz, 1990). Jane Drew was the wife of Maxwell Fry, and at the beginning Wendy Foster was an equal partner with Norman Foster. (Bristol Womens Architects Group, *Newsletter*, October 1991). See-sawing: in Bristol we have now started a South West Women in Property group. The committee members include architects Angela Crofts (a prime mover) and Christine Cuthbert King; Wendy Pollard, from a quantity surveying background, and major catalyst; planners Jill Jamieson, first woman chair of the RTPI South West Branch, Tracey Merrett, planning lawyer, and Lydia Lambert, local planner (and myself).

MANIFESTATIONS OF PLANNING FOR WOMEN

The generation of women and planning

There are a range of factors which generated the present planning for women movement, as explained by Judith Taylor (1990). The information given in this and the first section of the following chapter is based on Barnes (1989), J. Taylor (1990), B. Taylor (1988) and a range of contacts with other women in planning and specific local authorities replies and documents (some confidential). It would seem that a concern for minorities allowed minority issues to get on the agenda, for example Taylor (1990) argues that the effect of disabled persons legislation enabled social issues to become material considerations in determining planning applications or formulating development plan policy, because of the spatial design implications. In the early 1970s legislation had been introduced granting women maternity leave and also a form of equal rights, reflecting international trends in the wake of the second wave of feminism (Duchen, 1992b: 18), and the efforts of women Members of Parliament, few though they be (such as Audrey Wise, mother of Valerie of the GLC, see below). Change started outside the planning profession; for example, community groups concerned with childcare investigated the use of the planning system to achieve their objectives. Judith Taylor (and everyone else) mentions the SOCATECH initiative (South Canning Town and Caxton House) in the London Borough of Newham, where 9 per cent of the population are under 5 but no childcare was provided. This achieved planning agreements for nursery provision, which began to shift the planning agenda across from the public to the private side of the eternal dichotomy. Women were involved in public participation meetings in conjunction with the production of the new GLDP (Greater London Development Plan) in the early 1980s (Taylor, 1985), women being more involved than men in community groups as against formal politics.

Greater London Council

Barnes (1989) comments that in 1981 the GLC election brought in a form of local government, under the new left, headed by Ken Livingstone, completely different from the previous Labour group of 1973–7, who adhered to the Morrisonian tradition (of managing for the people). In 1982 a women's committee was established under Valerie Wise (Webster, 1983). This initial committee included eleven councillors of whom eight were female, and nine women co-optees of whom one was lesbian, one disabled and four black. They started with a budget of £350,000 and ended with £16 million before abolition in 1985. Barnes (1989) notes that their style of decision-making was that of informal Weberian democracy which reflected the involvement of many of its members in community politics and feminist groups. Although not a planning committee, they prioritized transport and planning, receiving support from some of the male planning officers. The new draft GLDP included a chapter specifically on women. The GLC also adopted a more women-friendly approach to the plan-making process itself. This was backed up by staff training in gender issues, and promotion of employment opportunities for women and childcare provision, and publications on women's issues, such as *Changing Places* (GLC, 1986a), which owes much to Jane Foulsham, and also to Jos Boys (and the PLC team).

London boroughs

The 1985 Local Government Act resulted in the GLC being abolished in 1986, the planning functions going to the London boroughs, with much argument as to the legal status of the new GLDP. But the influence of GLC policies spread throughout the London boroughs and to other cities. Both Tory and traditional Labour politicians were alarmed at this spread of ideas. Barnes (1989) notes that Labour leader Neil Kinnock's press secretary commented in the *Guardian* (7 March 1987), 'it is more anguish than panic, the London effect is now very noticable. The Loony Labour left is taking its toll, the gay and the lesbian issue is costing us dear among the pensioners', alluding to fear of extremism and higher taxes and rates in the GLC area, whilst one GLC Labour councillor criticized women's initiatives as 'absurdities' (Button, 1984: 74). Some Labour London boroughs were already well on the way to implementing women and planning policies such as Haringey, which was the first to appoint a special woman to deal with gender dimension of planning work, and already had a thriving women's committee (Poulton and Hunt, 1986; Button, 1984; Webster, 1983). Women planners themselves had banded together early on in the London Women and Planning Group and subsequently produced an extensively used document summarizing the women and planning policies found in each London borough UDP (LWPG, 1991) a document which one senior male planner, significantly, called 'exotic' (p.19). Women planners such as Sulez Takmaz, Ruth Cadbury and June Jackson (and many more) have taken a leading role in promoting women's policies within the context of the new borough unitary planning system introduced by the 1985 Act, and there has been considerable activity within other South-eastern authorities, such as the work of Colette Blackett and colleagues in Crawley and in Sussex. But, Taylor (1990) found that 20 per cent of UDP boroughs said they would not include separate women's policies (cf. Davies, 1993). Many did not make it through the Inquiry stage.

Other cities

In Sheffield the development of women and planning issues has been more gradual (Barnes, 1989). The Labour left came to power in 1981, but it was not until 1983 that an equal opportunities policy was established. A women's committee was set up in 1986, partly thanks to NALGO, who were concerned with childcare issues, and also male planning officers such as John Bennington were supportive. Dory Reeves has been active in the region, and in developing a forthcoming RTPI PAN (Planning Advice Note) on women and planning along with Susan Gudjonsson and Christine Booth. Christine first lobbied for change after attending the 1982 Polytechnic of Central London women and planning conference, saying it changed her life (*Planning*, No.479: 8–9, July 1982). (The power of the one-day national women and planning event cannot be over-estimated as a strategy for change: if only the men would come too.) The appointment of a women's employment officer helped to create the right conditions for change in Sheffield.

In 1984 in Birmingham the first women's unit was established (Taylor, 1990), partly because of the activities of the pressure group Birmingham for People, resulting in the report *Women at the Centre* (1989) in association with the Cadbury Trust and the City Council's women's unit. An early initiative was on transport, resulting a 'Women and Transport' report in 1989. The planning authority produced in 1991 'Women, safety and design in city centres'. It has been suggested that Birmingham adopted an *ad hoc* approach, but at least the Birmingham Revised Draft Unitary Development Plan (as at 1991) includes some women's issues, albeit under the section, 2.35, 'Equity/ Deprivation Issues'. In 1993 the 'Birmingham for People' women's group produced the video *Positive Planning* on women and planning.

Manchester established its women and planning group in 1985, and produced the excellent *Planning a Safer Environment for Women* (1987). Newcastle, although it contains one of the leading planning schools, appeared slow at first, but in 1988 a policy of providing childcare facilities for public buildings was introduced, and the UDP preparation programme included the activities of a group of ten officers working on 'disadvantaged people'. But it would seem that in regions where there is a history of urban deprivation, or of heavy industrial employment and docks, women's issues get blurred and marginalized, beside generalized social issues, and there may be a somewhat macho image to local policy-making: inner-city regeneration and ethnic issues being seen from a male perspective. Some cities shine out as being especially aware, including Doncaster, where Sunethra Mendis has been a leading light; Rochdale, which has always had a tradition of community planning (going back to the days of the Deeplish general improvement area project in the 1960s); and Leicester, which produced a joint police and planning working party report concerning urban safety by John Dean (1990). Once the London factor had peaked it would seem that the women and planning movement moved north (Calder *et al.*, 1993), whilst large cities in Scotland and Ireland have remained strong (Barrett *et al.*, 1991). But a few south coast areas come to mind, including Sussex (including Crawley and Brighton where there are many keen women planners) and further along, Southampton (1991), which has produced some remarkable work with Margot Duncalfe having a major input. Bristol, once seen by many as being over-preoccupied with urban conservation and traffic issues without always reflecting on the gender (and class) implications of their policies, in 1993 appointed a woman chief planning office: Diana Kershaw.

Internationally there is an increasing range of women who have made it to very senior planning positions, such as Dina Rachewsky, Israel's Director of national and regional planning, or Valerie Jarrett, Chicago's Director of Planning and Development who is a lawyer, black and young (36 years of age). Nationally, it is an ever-changing situation with some places starting strong and fizzling out with unfulfilled policies, and others starting slow and still building up steam. Whereas in some cases radical groups have given the system a jolt and got things going, in others there has been a stately, femocratic, build up, dependent on keen practical individuals: planning officers or women councillors, who, irrespective of political complexion, plod on. Nationally the RTPI is producing a PAN on Women and Planning as stated, and also an EO Policy Statement, written by Una Somerville (Belfast), Sandra Newton (Coventry) and myself in conjunction with the RTP Women's panel.

11

PLANNING FOR WOMEN

CHANGING THE AGENDA

Many of the problems which women encounter in the city of man are the result of a dichotomized public/private view of reality, prevalent within the planning subculture. In order to plan for women, physical divisions between perceived public and private realms manifested in land-use patterns must be dissolved. The nature of land-uses must be reconceptualized, and the likely inter-relationships among them reconsidered, to reflect more realistically the way in which women 'uze' urban space (p.22). In this chapter, I take a more prescriptive stance. First I describe ways in which progressive local authorities have sought to plan differently for women, illustrating this with a range of recent examples. Thus they have acted as zone zappers, disrupting the reproduction over space of gendered dichotomies and roles. Second I discuss ways of reconceptualizing city structure to break down spatial divisions to the advantage of women, by mixing, melling, and making new interconnections between, and clusterings among land-uses and activities; thus creating new spaces and possibilities for women. In this process new emphases, foci and zones emerge, centred on activities such as childcare and running a home, thus breaking down the physical/social (spatial/aspatial) dichotomy in the process, which has so constrained the nature of planning policy. Such activities have previously been relegated to the private zone, or fallen down the gap between the public/private realms, and been seen as other, or marginal and limited, but may yet be reborn as new land-use zones in their own right. Third I consider the legal means of effecting change through the planning system. It seems to me there is considerable intransigence, and ignorance, amongst some planners, but one does not have to wait until such persons change their minds towards, or agree with, women and planning policies in order to implement them, provided one can legally enforce women-benefiting policies through regulatory codes. Last, I suggest more fundamental changes in society. It is much more difficult, and beyond the scope of town planning, to change the patriachal perceptions and related professional cultures which generate dichotomies in the first place, but I believe one can exert positive influence through planning.

PLANNING DIFFERENTLY

Perusing women and planning policies (LWPG, 1991), it is clear that in structuring town planning women are likely to use a different set of classifications from men in organizing the problem. The document *Changing Places* (GLC, 1986a) uses categories of planning policy which are not land-uses in the conventional sense, but rather activities and topics relating to the way people uze the city, creating new linkages. These include: caring, housing, jobs and skills, protection of local shopping centres, safety from violence, and mobility. Mobility, for example, is seen as more than transportation; it is also walking (cf. Dean, 1990). Mobility brings with it a cluster of related issues such as safety, lighting, public transport, availability of public conveniences, and shelter, as discussed in *Women on the Move* (GLC, 1985). Disability is frequently brought centre stage within the context of mobility and access (Hillingdon, Policy BPS11 in LWPG, 1991: 6).[1] This is likely to be accompanied by a concern for employment opportunities for women (and men) involving the provision of childcare, and of safer, more user-friendly work environments. A local authority cannot seriously state that it is promoting equal opportunities within its boundaries, if it does not make childcare an integral part of its economic strategy; that is it deals with issues on the so-called private as well as the public side of the dichotomy. Likewise, ignoring the racial aspects of employment policy in locating new development, could be seen as contravening equality policies. But old attitudes still prevail. A local authority in the Midlands undertook enveloping schemes (that is improving housing and the immediate residential environment) in a black area, but they put the work out to tender and chose white contractors. Unemployed black people had to watch outsider white workmen improve their houses, when what they wanted was jobs, money and control of their own environment.

Sheffield has been strong on making links between the needs of minorities and employment policy. The Central Area Local Plan (Sheffield, 1986) identifies five policy areas which were more extensive in their scope than conventional spatial planning matters, namely to: develop training centres, improve employment and provide projects for women and other disadvantaged groups, encourage workplace nurseries, provide childcare facilities in association with shops, facilitate recreation and leisure for women too, and emphasize safety. The RTPI women and planning working party (1989b) stressed the importance of providing choice and opportunity for both planners and planned by means of employment and training policy, with emphasis on part-time as well as full-time alternatives. Part-time workers might be seen as prototype zone zappers whose daily work takes place both in the home and the so-called workplace, thus spanning the public/private divide. But they are often poorly paid for their pioneering efforts. Collecting data on both men and women, full-time and part-time, so the latter are not rendered invisible, is vital. Disaggregation (and recombination) of data on the basis of ethnicity, disability, location and other characteristics is possible nowadays with sophisticated computer software. In the public realm quantitative data seems to be seen more than qualitative material, for example through doing a social audit as in Newham in 1984 as a means of developing policy for the central area plan; and of Bristol city council staffing in 1993.

The women and planning movement has pioneered alternative ways of undertaking the planmaking process. Rather than adopting a top-down approach of planning for the people, the emphasis has been upon a participatory model of planning (Darke, 1990: 165–89), and upon dealing with immediate concrete problems, rather than adopting long-term, high-level, abstract strategies. Women planners, to a degree, are more likely to identify with the planned than the planners (McDowell, 1983), and may have their roots in community politics and so be more familiar with informal, qualitative, and serendipitous approaches to finding out what people want. Emphasis is frequently put upon more human issues, such as changing the image of bad areas by 'talking up areas' as Diana Kershaw, Bristol City's chief planning officer, puts it. Urban therapy, to change the spirit of the place and the life chances of individuals, is also a popular concept. Women planners have experimented with different ways of eliciting public participation; their methods might be renamed private participation as the emphasis is upon reaching women in their own territory, on women's side of the public/private dichotomy, in the shops, schools, and gathering places, rather than in public halls at mainstream meetings. For example in Sheffield a women's planning advisory service with a crèche on Saturday mornings was organized. The planners produced a 15-minute video and gave out 25,000 free newspapers in 1986 as part of 'A City Centre for People' initiative (Barnes, 1989). Planning education can also form a link with the community, especially when students are part of rather than separate from the community. Women's groups such as MATRIX (1984), and WDS (1993), and especially, Jos Boys, Sue Cavanagh and Vron Ware, within these groups, have done much to reconceptualize built environment education. Positive efforts have been made by South Bank University to attract more local young black people, especially women, onto planning courses. Women such as Patricia Roberts (previously Dean of the Faculty of the Built Environment), Marjorie Bulos and Maureen Farrish were instrumental in achieving this. The Newham women's working party asked a hundred schoolgirls aged 14–16 for their views, working in small groups. Much effort has also been put into explaining the planning system to women so that they understand what is going on, especially by voluntary planning aid groups (LPAS, 1986b) with women such as Ruth Cadbury and Catherine McBride being actively involved in advising women on planning issues. In London some women may be unwilling to go to evening meetings, if they have to walk across dangerous housing estates in the dark to reach some community hall.

Efforts have been made to reach black women. In South London, planning leaflets were left in Afro-Caribbean hairdressers. More black women are nowadays involved in housing management (Papafio, 1991; Rao, 1990, WDS, 1993) but fewer in architecture and planning. If the planners have previously shown themselves racist in operating the development control system, black people will not trust them. Accounts abound such as about a black women's centre in East London being required to have a low noise level by the planners; of requirements for unnecessary screening and planting in front of mosques, excessive car parking standards being stipulated for an Asian women's centre in Hounslow, and even of a pentecostal church being refused permission because of concerns about flooding because of the practice of total immersion baptism (cf. Thomas and Krishnarayan, 1993; RTPI, 1993). Male planners may consult with the Sports Council, the National Playing Fields Association, road user bodies, male politicians, and workforce representatives, and with their colleagues in transportation or in the city engineers and surveyors departments. Women are more likely to be in contact with community groups, including those concerned with childcare, public transport use, and consumer issues. Their internal contacts within the local authority are more likely to be with the education department, health bodies, housing authorities, and the social services, leading to new policy linkages such as providing childcare in conjunction with women's skills centres (Deptford), and linking local planning to community minibus organization (Southwark). In the London Boroughs Disability Resource Team issues of disability, gender and race are dealt with together by the same people in respect of planning applications. Women may encounter opposition from male groups within minorities who may want more sport rather than childcare provision. White male planners appear keen to oblige, because they see sport as the panacea for all social ills. Developers may offer planning gain in the form of male sports facilities when seeking planning permission for new retail development whose main users are women shoppers. Black arts centres, and dance centres (Bristol), as against fitness training centres, are generally given low priority.

If women's policies are put in a separate chapter it gives space for their full development, but may give the impression to zone-prone planners that women's issues are other. If they are integrated within each topic (such transportation, housing) they may get lost, but direct linkages can be made as to the implications for mainstream planning. No one talks about separate men's planning policies, and some women planners prefer to talk about planning for the majority (52 per cent) rather than planning for women. You can never plan entirely separately for women (or for men), and it is a sign of dichotomized thinking (and pillorization) to imagine so. We all live in the same urban space, and women's planning policies are not aimed at building a separate women's super-city elsewhere, which drains resources away from men's needs. Many women planners argue that if you plan well for the minorities, then the majority groups also benefit from the higher level of public provision and improved environmental conditions. Indeed planning for women inevitably means altering planning for men and may be seen as an all-inclusive approach which affects everyone and every aspect of urban form and development, and not as a limited, marginal little policy option for a small group in society. As proponents of the systems view of planning stressed (McLoughlin, 1969) everything is linked to everything else. If you change a land-use here or a transport route there the effect will reverberate right through the system, and have aspatial (social) implications for everyone. Examples of the integrated and separate approaches are given in *Shaping our Borough* (LWPG, 1991), which includes exemplar UDP policy statements on most topics. Lambeth originally had a separate chapter on women, Hillingdon 1987 added a special chapter on disadvantaged groups, and yet Merton, Newham, Lewisham and Deptford had women and planning threaded all the way through (cf. Davies, 1993).[1]

PRIORITIZING SAFETY

Over and above spatial concerns, the aspatial issue of safety features strongly in many women and planning policies, both in relation to transport issues, and as a component of policies on different types of land-uses and developments. I have singled this issue out because demands for assurance of private personal safety for women entering public space often appear to be foremost in the minds of ordinary women when they consider what planning can do for them. The issue of safety seems to be located right along the earthquake fault line where the public/private dichotomy interfaces. This is evidenced in studies undertaken by the Suzy Lamplugh Trust (Lamplugh, 1988), and in a survey undertaken by Southampton planners (1987) on the safety of women in public places. This study was based on a broadly ethnographic approach which, arguably, gave a far better picture of women's experiences than crime statistics alone. Several other authorities have produced design guides on safety, including Waltham Forest's report in 1988 entitled *Access for All*, Southwark's *Housing Security Design Guide*, and Manchester (1987). Other authorities such as Birmingham (in April 1991) have produced internal reports on the situation to inform planning officers. Safety policies may emanate from highways departments, seeking to reconcile aspects of road safety with woman safety which have often appeared at odds (Davis, 1993). Doncaster, Sheffield and Rochdale now have policies against the perpetuation of pedestrian underpasses, thus putting safety before traditional car transport priorities. A woman highways engineer working in Lothian, Scotland, was responsible for getting street lights changed from yellow to white, which improved safety for pedestrians, and Haringey was the subject of a campaign to improve street lighting.

Many women do most of their travelling by foot and are dependent on public transport. Some authorities advocate no more than 250m as a reasonable walking distance to local facilities or bus stops (LWPG, 1991: 17). In the old days, when men walked too, 200 yds or 3 minutes' walk, was seen as an ideal figure (p.131). Adequate lighting is also advocated for pedestrians, as well as seats in shopping centres, ramps for buggies and the disabled, public conveniences, and minimal steps and unnecessary changes of level or surface texture. To increase safety the use of boundary railings (Ealing), or pierced walls using spaced brickwork rather than solid walls is recommended, particularly around entrances (for example to parks or housing units) so people have a clear line of visibility. Making public transport more accessible and safer for users are high priorities, with British Rail coming in for special investigation (and much condemnation), but everyone mentioning a successful campaign for a footbridge improvement incorporating security fencing at Honor Oak in Smethwick in 1986 (Taylor, 1990). A survey of British Rail and underground stations in Hackney (CILT, 1990) made visible many problem issues, typical of the whole London system. These design issues are not, strictly, town planning, but they are relevant to quality of service, and rights under the Citizen's Charter.

Southwark planners (LWPG, 1991: 12, Policy Env. 1.1) require that new public buildings should be designed to overlook public areas, and that use of public areas should be increased by the encouragement of mixed land-uses. Many authorities are concerned about dead city centres in the evenings. It is being suggested that traffic should be allowed back through pedestrianized shopping streets in the evening, and empty property above shops should be reused for housing. Unless the entrances to such property are well lit, and preferably at the front, many people would be loathe to live in such flats, if they have to walk through back service areas of shops to their door in the evenings. Car parks are another source of potential danger, but it is illegal to discriminate against men by providing women-only car parks (see *Daily Mail*, 27 June 1991: 32, article by Jessica Davies, 'Why can't we park in safety?' as a reflection of public concern). Separate areas might increase the problem, the Automobile Association recommending more attendants overall. Underground and multi-storey car parks are seen as threatening edifices by many would-be users. Good lighting, security, attendants, and better design would improve matters, but why not also public conveniences, covered bicycle racks, cafés, crèches, left-luggage areas, and dog-minding areas (the facilities being available to anyone, not just car drivers) paid for out of parking charges which are already very high. The car park might take on the role of an activity nucleus.

DEZONING

Home and work

It is essential to restructure and integrate land-uses at city-wide level, to break down impractical land-use dichotomies to the benefit of women (and thus all members of society) (Barrett *et al.*, 1991). Many women and planning policies start with trying to undo what has gone before, especially the effects of corrective zoning. Reintegrating work and home raises questions about the role of the residential area. A consequence of women going out to work is that residential areas are expected to accommodate nursery and childcare facilities, which interestingly are defined as Class DI (b) non-residential within the Use Classes Order (Moore, 1987: 305), and thus require planning permission. Class D seems to be an 'other' use category generally, also including religious buildings and aspects of leisure and cultural provision; all safely separated from the serious uses of business and industry. Permission is not normally needed for the use of a room within a domestic house for childminding unless the use becomes dominant or intrusive, or, in some cases, if the dwelling is part of a block of flats. There are several design guides on setting up commercial daycare nurseries (Greenwich, 1991), and many planners try to be accommodating in dealing with such applications. But some authorities still refuse new nurseries because of alleged inadequate car parking provision, which is ridiculous when many women (and all children) do not possess cars. Matters are not helped by the unwillingness of some local authorities to register new childminders (or by taking two years to do so). The 1989 Children Act requires childcare facilities, even in people's homes, to conform to manmade space standards, which is ironic considering there are few space standards imposed on the internal dimensions of rooms in private housing (except for building regulation controls). As Carmen Duncan (Scottish branch and Equal Opportunities race panel) commented, in conversation, there is a case for greater concern for internal as well as external design controls, because some dwellings are so very small, echoing the long standing insides/outsides debate, *provided*, women planners can set the standards and control the agenda.

Within residential areas, I would like to see the adaptation of existing buildings, along the lines suggested by Hayden (1984). She suggests that existing neighbourhoods could be modified so that, say, one house in twenty was converted to a childcare centre. This centre could be also provide other useful facilities and services, such as somewhere to leave pets, elderly relatives, and parcels in a crisis, with possibly a café and laundry service provided too. This is only a half-way house towards the full blown co-operative house-keeping ideas of the last century, but we have to start where we are at and work our way forward (or possibly back). Childcare provision is totally inadequate in this country, and more government investment is needed, possibly in the long term in the form of purpose-built facilities, in new care zones midway between employment and home areas. Since we have an ageing population, reduced fertility rates, and an emphasis on choice, it well may be that some areas would not be designed around childcare to the same extent, but rather around care in the community, or a collegiate system of single apartments with the choice of shared restaurants and meeting buildings (either mixed or single sex). In all cases developers would be required to come to a planning gain agreement for community provision (Grant, 1982; 292–4).

Trying to remix zones is not easy, as it is 'not what people are used to'. An alternative childcare solution is to create workplace nurseries. Taking children through the rush hour to a central work place crèche is a controversial issue. In the same way that one car space may be provided for every 200 square feet of office space, why cannot one crèche space be provided for, say, every 500 square feet? Increasing or restricting such standards, as with car parking, could be an additional tool for controlling central area development, and would transform the appearance and mix of central areas. Some developers are realising that the provision of childcare in out-of-town business parks is a compatible use which will attract more women workers (for example at Waterside Office Park, Northampton). Some planners see crèches as an acceptable component of the mixed B1 Business Use Class (1987 Use Classes Order) (LPAS, 1986a and b). This could be done easily on industrial estates, as in the war.

The opposite – bringing the work home – has received a mixed reception. An RTPI distance learning unit (Thomas, 1988: 1–25), takes a negative view of such activities. But Southwark's draft district plan (Policy E.1.7, LWPG, 1991: 15) recommends that such uses should be permitted provided home-working is limited to residents of the area. Whilst many women, some men (including employers who welcome money-saving flexible working ideas) see home-working as the future ideal, with everyone sitting at home hunched over their computer with a modem link to their office, some people declare, 'I can't work at home, I have to escape to the office'. Unless there is a shift in gender roles and responsibilities, moving work to the home, or childcare to the workplace, is only half way there. Changes are also needed as to legal conceptions of property tenure and ownership so that women have more access to housing and control over their environment (RTPI, 1989b: 116). Tenure (Circular 7/91, updated by PPG 3, Housing, D.o.E, 1992c) is now a material planning consideration in order to provide affordable housing (but not on gender grounds).

If erstwhile inner-city ethnic minority areas are gentrified and declared conservation areas, residents may be forced out because of design controls restricting the external appearance of the building which would prevent them making internal divisions. Orthodox Jewish families ideally require two kitchens, if they can afford it, a milky and a meaty kitchen because of kosher food preparation law. This may involve a house extension (cf. GLC, 1986c in relation to Haringey's Tottenham area). Moslem households may operate a gender segregation of space within the house, with a separate *zenana* area for the women (p.80), which only works well when there is adequate space to start with, and some sort of separate internal courtyard room. Who are we to judge the political correctness of planning for separate women's quarters, for, without them (like some of their Anglo Saxon sisters), women may end up worse off, stuck in the kitchen whilst the men take the public living room. Such matters must be treated tolerantly, even when such traditions perpetuate gendered roles and zoning divisions, as planning refusal may indirectly contribute to the breakup of communities. The situation is complex.

Redefining public open space

I am featuring this component of urban form because planners have traditionally prioritized the importance of space for sport. As explained (p.117) planning theorists have long been concerned with accommodating three elements: home, work and play in the city of man (Doxiadis, 1968). A problem for women is that men's field games, children's play, and even childcare have all been elided and merged as the same activity – sport – to legitimate vast amounts of land being used for public open space. But sport is not for all, for example women are not mentioned in recent policy guidance on sport and recreation (D.o.E, 1992d). Parks in large cities, especially London, have traditionally provided a detailed range of facilities, especially for (so-called) passive leisure (Thurston, 1974; and Whitaker and Browne, 1971). Women have stressed the qualitative aspects of open space, rather than quantifiable standards. Caroline Hanson and Jacqueline Burgess (*The Planner*, November 1988: 127) stress the detailed features that people like (such as parks with pools and ponds) and include a photograph of two women paddling in a pool with their children. This is compared (quoting from a Royal Society paper) with the contemptuous attitude some planners exhibit towards the masses, who are seen as ignorant and in need of control so they don't mess up the countryside too. Many people do not undertake informal recreation in the countryside, including many who are female, elderly, disabled, from ethnic minorities, single parent families, and many children. However, the deciding factor determining access may be car ownership rather than belonging to a minority (cf. GLC, 1986b). An emphasis in many women and planning policies is upon disaggregation, precision, and detailed consideration of who exactly needs what, with a greater emphasis being given to local facilities, and appreciation of the real space requirements of children's play. Anthony Fyson observes play needs space (*The Planner*, 21 September 1989: 3). The Dublin group (Barrett *et al.*, 1991: 38–46) identifies, with sensitivity, the different types of open space areas, and categories of land-use within public parks.

The men could keep some of their playing fields, and women could have more of them too. But I would prioritize the development of new Victorian style parks with duck ponds, flowerbeds, seats, safe supervised play areas and plenty of public conveniences (cf. WDS, 1990). I would have distinct dog-free areas, but also extensive areas for those who wished to exercise their dogs. Otherwise women dog owners might be pushed into using more remote, dangerous areas. In Switzerland there are apparently separate dog parks already. There would also be far more space for allotments within towns. As to the countryside, there would be no factory farming permitted, which would result in a more interesting countryside and happier animals. With the blurring of town/country dichotomies, and discouragement of restrictive covenants against the keeping of pigs and poultry in residential areas, it is likely that one would also find more animals and chickens wandering around urban neighbourhoods, thus incorporating some of the better features of so-called third-world cities. Regarding clean/dirty dichotomies, definitions of hygiene might change. Currently dogs may see humans as the dirty ones polluting their favourite beaches with sewage.

In discussions on integrating work and home, and childcare location, some women planners have commented that there is nothing more beautiful than the sound of children playing. Lesser mortals have expressed concern about noise from (other people's) children. Further, play streets and integrated play areas alongside houses have always received a mixed reception from residents, especially from people working night shifts, those with babies, and those working at home. Standards such as the provision of children's play areas within 400 metres of all dwellings (Birmingham Draft UDP, page 21, para. 3.47) are still to be found, which reflect a traditional quantitative rather than a more sensitive qualitative approach to the issue. In some schemes the developers have donated the left-over bits of landlocked space between the house plots as play areas because they are seen as no use for anything worthwhile. There is much to be said for the practice in some (erstwhile) socialist states of locating both schools and palaces of sport near to employment rather than residential areas. Provision of adequate programmes of after-school and holiday supervision and care is required, with special amenities for older children and teenagers linked to school and sports buildings – with the choice of separate (and mixed) facilities for girls and boys. Amenities could also include private areas for study, and quiet, possibly linked to public library and local museum space, arts, music, and crafts facilities, and computer terminals. This could develop into a distinct care zone alongside (or between) housing and employment areas, linking to truly public open space, and acting as a cross-over zone between the public and private realms.

Shopping and food gathering

I would replace play with shopping as the third most important element, after home and work, because it is food gathering for survival, and in order to enhance the status of so-called consumption which should be seen as a form of work (like production). However, I am aware that this might also reinforce stereotypes of it being a role which women do as a form of play or leisure, as a result of being selfish or over-materialistic, when in fact they are usually doing shopping for others, not for their own enjoyment. See-sawing: in one of my tutorials, when discussing the need for shops to be near housing estates, one of the male students shouted out, 'you mean so that the wife can take her husband's credit card and waste all his money on buying clothes', interrupting one of the female students who was struggling to present her concern about accessibility to food stores.

Shopping is the basis of an important bridging and reconciling middle zone, where public and private realms meet: literally across the counter. But this does not mean they have to be all in one clump, engendering unnecessarily high land values (in the central area) or access problems (in out of town centres). Many women and planning policies stress the importance of maintaining the existing hierarchy of shopping centres to protect local shops (Hillingdon, LWPG, 1991: 7, Policy, BP S24). Evening use of shopping centres, not only breaks down the legislative division between public and private time (reflecting in public opening hours), but is more convenient for women provided there is adequate public transport and lighting (and public lavatories do not close at 6 p.m.) – and provided that much ignored shops premises legislation which entitles assistants to have access to a chair is enforced, so they do not have to stand up all day and night.

Adequate provision should be made for women with children, both as customers and staff. Property professionals are catching on to this; an article in the the *Valuer* (Vol. 60, No.8: 1, and 16–17, October 1991 'Creche course for developers') states mothers spend more if unencumbered by children. Lewisham has produced a guide to facilities for customers (Taylor, 1990) for developers of large new stores with an emphasis on toilets and crèches. Kensington and Chelsea (LWPG, 1991: 13, Policy 5.9) stipulates adequate distribution of public conveniences throughout the borough (empirically I find this is not the case), which includes part of the main tourist centre of London. (Why, incidentally, are abled toilets not shown on the A-Z map of London?) Certain categories of buildings are already required to provide toilet facilities, although compliance is poor. Categories include large stores, restaurants and, suprisingly, betting offices. See-sawing, the new Galleries shopping development in Bristol is designed with the toilets on the top floor, with less provision for women than men (although women are 80 per cent of shoppers) and with heavy, awkward double swing doors at the entrance to the Ladies. £6 million was spent on refurbishing a shopping centre in Cardiff and still it does not have door openings wide enough to take a double buggy. (Planning actually seems to have regressed in this respect, from the days of the old gentlemen planners, such as Adshead in the 1930s, who stipulated 5-foot widths to accommodate prams, p.133.) Many design guides stress the importance of baby changing facilities and breast-feeding space, but few developers heed such guides, and yet women can be accused of disturbing the peace by the police if they carry out these necessary tasks in public.

The caring city

In seeking to facilitate activities such as caring, within a safe environment in which women have choice and opportunity for employment, mobility, and personal development, with adequate childcare, shopping, and leisure provision, certain policy objectives are identified repeatedly in the various documents. These are, first, to reduce the need to travel, by increasing the mix of uses in areas (especially reintegrating work and home), and second, to provide a greater distribution of localized, small-scale, friendly, safe facilities, shops and amenities: effectively a multi-nucleated city. Third, women want better public transport, so we do not have that feeling of being stranded, and this is equally important for those living in the countryside. Finally, women's planning policies are generally against out-of-town shopping centres, dispersed housing development and distant suburbs: that is, land-use patterns based on the motorcar. Centralized (or concentrated decentralized) provision of facilities such as hospitals, comprehensive schools, and government offices to achieve economies of scale (for whom?) which diminish the provision of local facilities are to be discouraged. A return to a traditional hierarchy of centres and shopping provision would be welcomed, as most women are not anti-urban and want to retain the cultural benefits of central areas as well as having the convenience of local centres.[2]

PLANNING LAW: PUTTING POLICY INTO PRACTICE

To enforce women and planning policy one has to step into the public-realm of state planning law. To make developers provide facilities and modify designs one must use planning gain agreements, or impose conditions on the planning permission; otherwise we are just talking ideas and dreams, not implementation and development. There are possibilities and limitations inherent in planning law. Any conditions put on a planning permission must be for a genuine planning reason (Morgan and Nott, 1988: 139), this judgment being highly gendered. Some men planners still consider women and planning issues *ultra vires*, stating, 'women – that's not a land-use matter', but women argue that gender *is* a material consideration in planning, because women and men use space in different ways, (Taylor, 1990: 98). In respect of the unitary development plan system, Circular 22/84 (updated by PPG 12, 1992 concerning development plans) states that the unitary development plan system will provide authorities with positive opportunities to reassess the needs of their areas, resolve conflicting demands, consider new ideas and bring forward appropriate solutions. This has offered a foot in the door to women's issues. In granting permission, planning authorities can impose such conditions such as they think fit. This was previously embodied in Section 29 (1) of the 1971 Town and Country Planning Act. Further guidance as to what is considered valid, is given in Circular 1985/1, 'The use of conditions in planning permissions' (and see PPG 1, 'General policy and principles', D.o.E, 1991). Also, the local planning authority is required by law to take gender issues into account in the provision of public facilities under the 1975 Sex Discrimination Act, in so far as it is illegal to refuse or deliberately omit to provide goods and services because of the recipient's sex. Similar requirements apply under Sections 19 and 20 of the 1976 Race Relations Act. Section 5.48 of PPG 12 (Planning Policy Guidance note) on local plans requires local authorities to consider the impact of policies on ethnic minorities, disadvantaged and deprived people. Provision of access for the disabled in determining a planning application is a requirement of Section 76 of the 1990 Town and Country Planning Act, and the Code of Practice for Access of the Disabled to Buildings (BS 5810: 1979) gives specifications. But certain authorities are notorious for refusing single-storey back extensions or widened front door porches to accommodate wheelchair usage; as if they expect the house extensions to all go somewhere else outside their area. As to access to planning education, Design Note 18, *Access for Disabled People to Educational Buildings*, published in 1979 by the D.o.E gives guidelines, although many colleges cite problems of compliance in older accommodation. The RTPI *Code of Professional Conduct*, 1986, makes it illegal to discriminate on the basis of race, sex, or creed, and religion. Many have found that application and enforcement of these regulations and principles has left much to be desired: some planners seem unaware they exist.

Planners have sought to get developers to provide or contribute to women and planning facilities by means of a planning agreement, that is by obtaining planning gain under Section 106 of the 1990 Town and Country Planning Act (amended as Sections 106A and 106B, by Section 12 of the 1991 Planning and Compensation Act in respect of planning obligations (unilateral agreements); and previously known as an S.52 agreement under the 1971 Town and Country Planning Act). Such agreements must only be used for the purpose of restricting or regulating the development or use of land and must be reasonable (Circular 22/83 and 16/91). Planning gain works best in high-value areas where developers are desperate to develop, and are willing to pay what many see as an additional land tax. Even if the developers are willing to build social facilities, someone has got to pay for their maintenance and management. Local authorities cannot afford to pay for this because of government cutbacks. A good example was the Section 52 agreement on the Ropemaker Street site, EC1, in the London Borough of Islington in 1985, where a crèche for thirty children under the age of 5 was provided to go with the new office block on the site, plus funding for ten years. Similar agreements were obtained at Orchard Square and Meadowhall in Sheffield covering crèches, toilets and baby changing facilities (Taylor, 1990). Other authorities have entered into agreements with developers to improve women's training (Newham), leisure facilities (Hammersmith and Fulham), and public transport (Camden Hopper bus service). Agreements must now be directly linked to the planning permission or they will fall foul of the 1989 Local Government and Housing Act, which requires 50 per cent of capital receipts to be set aside to repay current debts; this is a complex rate-capping issue that can limit the provision of social facilities (Aisbett, 1990). Lambeth has attempted to use the weapon of land ownership to provide social facilities, and compulsory purchase is also a significant planning power. However, local authorities have been under pressure to sell land in order to meet their financial deficit. The mainstream debate about the extent to which town planning should intervene in land ownership, and whether developers should be subject to a betterment levy to plough back some of their profits into the community by contributing towards physical and social infrastructure, continues within feminism.

Provided that clear policy statements as to what is expected are written into the relevant statutory plan, developers may reasonably be expected to provide related social facilities, without the need for complex planning agreements. The Department of the Environment appears to prefer this approach than leaving it to the *ad hoc* imposition of complicated conditions of permission at the planning permission stage, or to complex planning agreements. As accepted and approved policy, these women and planning statements have the force of law when determining the planning application, but they have to be durable enough to stand up to a planning appeal: the fact that they are good is not enough. In the final analysis what happens in a particular local authority depends on the attitudes of the local planners as to whether they are co-operative or negative in their support of such issues. There is a need for training of professional planning staff to be aware of the issues. The existence of senior women officers, or at least sympathetic male officers, and support from the councillors themselves are essential (RTPI, 1988), the existence of a women's committee being particularly useful. It is a matter of political will. For example, Barnes (1989) comments that Hackney allows for a flexible interpretation of the law in respect of change of use, and that Haringey tends to overlook technical contraventions because planning is seen as being about interpreting the law (especially on *sui generis* uses, and changes between use classes); it is not a mechanistic activity. Attention must also be given to monitoring of planning decisions to ensure that policies are properly implemented.

A GENDER CITY?

At the end of this book, sitting on the edge of the next millenium, I offer no ideal plan for a gender city to rival the garden city conceived a century ago (Howard, 1898). Indeed, presenting a master plan might be seen as a patriarchal approach. I have given an account of current women and planning policies and implementory procedures, with which I broadly concur. My emphasis is upon transforming existing cities, not starting again. This should be seen as a positive constructive attribute, not a sign of failure. I am not proposing a gender city because this approach marginalizes women (as, subconsciously, gender usually is taken only to apply to women). Rather, I am seeking to change all aspects of existing development to recreate the main-stream city in a manner which acknowledges the existence and needs of the other half of the human race, but in which men would live too, their lives integrally transformed. In this process the economic, political and social issues and problems upon which the malestream has focused (not least capitalism and employment) would also inevitably be fundamentally altered. I envisage the ideal urban form as a multi-nucleated city of cellular structure. I would like to see broken down the division between the suburbs and inner city areas (rather than between town and the country) giving each the advantages of the other. I would decentralize employment and commerce from the city centre into suburban areas, in order to provide them with a full, remedial, range of additional land-uses, job opportunities, and facilities to make them truly balanced communities for women and men. A detailed patchwork could be built up, with a diversity of local areas, all fully provided with local amenities and facilities. New housing development would be much more compact than at present with the option of private gardens around houses, but also medium-rise serviced apartments would be encouraged for those who wanted more back-up and no gardening. By the time we had fitted in the open space uses suggested and community facilities there would be no room for useless swathes of landscaping, but the city would be greener, and land values more even overall. Everyone would be in walking distance of at least some shops, amenities and local employment.

Such shifts in land-use patterns, by defusing the congestion of central areas, would solve many current transportation problems – of commuting, parking, and traffic congestion – as fewer people would make long journeys to work. But, there would be a need for a certain amount of traffic calming (an expertise developed by women such as Carmen Hass-Klau *et al.*, 1992), but the choice of car transport would still exist, especially in the interim before the new policies became more effective. Gradually the effects of present planning policies would recede, as the new policies took effect (but it could take several years), so that people would not *have* to drive to an out-of-town centre to do the shopping, or to a distant hospital in an emergency, or to get children to school. People might need to use their car to carry loads, to transport large families, to take an injured animal to the vet, and as a recreational vehicle in the countryside, or for longer trips. But, public transport would be improved beyond all recognition so this might become a less popular freedom. But not everyone can be expected to cycle or walk, as some macho-greenies imagine. The existence of so many local facilities and businesses would create a vast array of new jobs, and also more flexible forms of employment might incorporate working from home or from new purpose-built mini-office blocks within residential areas, and would be linked to wider changes in the construction of the economy itself. This is not a middle-class tele-cottaging electric dream, nor a petty bourgeoisie nation-of-shopkeepers scenario, because some work is still best done (at present) in a purpose-built structure altogether with other people, be it a factory, college, or office. But at least the pressure would be taken off certain areas by the dispersed multi-nucleated nature of development, and the land-uses throughout the city would be much more mixed and balanced overall, and thus commuter generated traffic jams would be reduced. Sophisticated, strategic planning at city-wide level would be needed to create this finely grained detailed city form and structure. But then so many more options would be permitted development than today that planners could concentrate on the more strategic, qualitative and pro-active aspects of planning – and on talking to people – and less on negative, controlling aspects.

One can only do so much through the statutory planning system and the above future scenario could only be achieved in concert with the operation of a range of other governmental powers and economic measures, plus a change in attitudes, gender roles, food production and distribution, and the functioning of business cultures. There is a need for a move towards more flexible ways of working which would enable women (and men) to order their lives more logically (RTPI, Yorkshire, 1991a and b), demands from women planners mirroring demands from working women in society. (I discussed the changes needed in education and practice for women in the built environment professions in the last chapter of my previous book, Greed, 1991.) At first, fiscal measures would need to be introduced to protect and subsidize useful shops, businesses and services in the local areas; controls would have to be put on big business, which is bound to argue that it could offer the economies of scale and be more efficient and profitable. In reply, I ask: for whom and at what social cost? The reconceptualization of work would not only make redundant the ancient dichotomized beliefs about the relative merits of production and consumption (words such as prosumption making a début, cf. McDowell, 1991) but would undermine the zoning of people into inappropriate social classes according to masculinist categories.

There are dangers in an apolitical vision of spatial transformation, for a post-modernist perspective has a habit of softening and blurring power differences and hierarchies, and turning oppression into a mere philosophical concept. Some women are wary that women's issues may just become one interesting factor among many, or subsumed under the broader category of gender. Whilst admittedly men are policed and controlled as much as women by the zoning and layout of the built environment, the current trend for replacing discussions of geography and gender with that of sexuality and space may detract from women's disadvantaged position, and deflect from the power of patriarchy. The shift from an emphasis on women's issues, to gender studies, to abstract discussions of sexuality, is a slippery downward slope in respect of distracting from the specific practical needs of women. On the other hand, removing the zones, or remixing the zones, does not negate the power structures which maintain patriarchal society, unless backed up by societal change. A phoney form of zone zapping may be created as superficial as social-mix policies in new towns (p. 000) unless underlying patriarchal attitudes are changed. Although an essential step towards change, some argue it is wrong to press for new zones which cater for traditional conceptions of women's work as childcarers and housekeepers, which would perpetuate male/female role divisions and sexist stereotyping (and contribute towards the anti-feminist backlash, cf. Faludi, 1992, and a rerun of the only half-way to Paradise scenario of fifty years ago (Wilson, 1980), when women were planned for as housewives and not as zone zappers) But, as stated, this argument is spurious; childcare is needed – whoever does it – and when gender roles change in the future, childcare facilities will be there for the men to use too. A more subtle negating argument I heard recently was, 'these issues affect everyone and so they are not the specific concern of the planning profession and should not be included in policy statements'.

CONCLUSION

In order to assess the chances of changing planning one must identify its power source. In my study of surveying, it seemed that capitalism and socialism were but two sides of the same patriarchal system which shaped the built environment (cf. Stewart, 1981). From this present study it seems to me that capitalism, socialism and patriarchy are but three sides (among several) of a polyhedron, all of which reflect aspects of the dichotomy-centred nature of dominant belief systems; the root cause for this, is, I believe, a theological issue, although patriarchy, and the related dichotomized world view, is a primary manifestation of the problem. It seems to me that traditional malestream criticisms of capitalists and capitalism as the culprits responsible for human inequality detract from the undoubted power of patriarchal, managerial structures in public-sector bodies, to keep women in their place. Such managers might redirect potential criticism of their own unwarranted power and high incomes within state bureaucracies onto the capitalists. The maintenance of patriarchal power and belief appears to be a driving power in itself as the spirit of planning, and yet is maintained, by its cultural manifestations in society, and in the design of the built environment, thus feeding back on itself. Local authority planners appear to have considerable power, but it is fragile, because it is, arguably, only a reflection from outside of other power sources, be they political, ideological, economic or religious. Town planning may be capable of endless metamorphoses, the professional arena acting as a place of discourse, a competitive ring, a hall of mirrors, a transmitter, for trends, ideas, and demands coming in from outside. It seemed to me from this research that planners are relatively weak, and easily swayed by the ideologies of others.

A frequent theme has been the power of the patterns of the past which reassert themselves in aspatial and spatial matters (cf. Foucault, 1972). We are, living in the overlap, seeking to implement new planning policies within old belief structures and established land-use patterns. Of necessity we seek transitional, sub-optimal solutions, to accommodate women's existing, oppressed role in society by immediate, practical measures, not least to reduce the oppression of rush-hour commuting. This mirrors the classic dilemma as to whether town planning policy should be opportunistic and incremental, or long-term, pure and high-level goal-orientated, (cf. Healey *et al.*, 1988). Many women believe that fundamental change can be brought about by small-seed initiatives, from the grassroots of society upwards (because the personal is political), as well aiming at larger societal shifts. But, of necessity, we have to be pragmatic. We have got to stand and face it, seeking to implement change by whatever means possible. Waiting for the revolution is no policy, and can be seen as a delaying tactic to avoid dealing with women's immediate needs. There is no North (for the slaves to escape to), or feminist Jerusalem (to go to next year), and no male genius or earthly Messiah to offer us salvation from the evils of the city: we have to create our own realities. I'm finishing this book on a downer, and, as was the case in *Surveying Sisters*, much of the content has been negative. Likewise, Ramazanoglu (1992) concludes pessimistically that we must face reality and admit our current limitations in effecting change, but, significantly, she still talks of imagining the future (1992: 287): rather than waiting for some male genius to do it for us. We can all imagine, and plan; it is not selfish, or likely to cause disruption as we *are* society, and the majority of the urban population. The target of our imagineering and amelioration of the urban condition must be existing cities, since it is unlikely in Britain that we can go off and build a new ideal city from scratch (North America is vast compared with Britain: Texas alone is four times the size of France, and six times that of England). One can, however, be inspired by an image of what it ought to be like, in the mind's eye, as something to aim at gradually. In dealing with matters such as control of the motorcar, and provision of public transport, one cannot put the clock back, or change everything immediately through draconian negative controls (and betray all those women and men who *have* to commute or cannot get between home → childminder → and work, early in the morning without a car), in the name of some ideal of the sophisticated, traffic-free, European city of the future.

The women and planning movement has restructured the discourse of planning – for women at least (sometimes it seems there are two planning professions). Although women may be seen as just being concerned with social issues, their policies are translated both into policies for restructuring city form, and into detailed, spatial standards on, for example, baby changing areas, shopping space, office development, play areas and public conveniences (Caffrey and Éanaigh, 1990; WDS, 1990). In this process women cut across the false physical/social dichotomy which has divided men planners throughout the twentieth-century, reuniting the public and the private, setting the groundwork for reuniting home and work, and dissolving the division in the process. Women are not simply concerned with producing detailed design standards in which form follows function, but with *force determining form* (the life force sustaining the human race, p.106) that is with creating whole, new ways of living. Women planners have a key role to play as philosopher queens, to produce alternative visions of the city, for as Anaïs Nin commented (1978) we write to create the world we live in. Creating cultural realities, and getting others to believe in them by our confidence and actions, and the good sense of our proposals, is the power which can shape the built environment.

As to future strategy, it seems to me that it is vital, by writing, through planning education, and by consciousness-raising within the profession, to make planners aware of the folly of a dichotomized world view, and to present them with persuasive, convincing, alternative urban realities: translated into clear policy guidance where possible. But even if senior men planners do agree that women have a valid point, and perhaps praise them for their efforts, it does not follow that they are going to do anything about it. The current need – ostensible need – for cut-backs in resources seems to be applied in a noticeably gendered manner, with plenty of money still being available for the creation of new managerial levels. In contrast, I have come across many examples of women whose women and planning activities are being squeezed out of their professional lives, because of the pressure being put upon them by their bosses in terms of increased workloads, hours and responsibilities. Women who were once optimistic, liberal-minded feminists, are saying 'we don't seem to be getting anywhere', and are becoming more radically minded. Some feel betrayed, and angry about what has subsequently proved to be the transient nature of interest shown by some ostensibly supportive male planners. There is a growing distrust of male allies who speak with a forked tongue; who encourage and put the brakes on at the same time; who 'face both ways to please everyone'. The movement may need to become more political, and self-sufficent in order to push for changes in legislation which make equal opportunities policies more enforceable.

It seems to me that mandatory codes are needed, to implement women and planning policies (for women as consumers of the built environment) that can specify standards and performance criteria which must be complied with in a development.[1] These might cover issues ranging from widths for doors in shopping centres and minimum requirements for provision of public conveniences (the insides as well as the outsides of buildings would be seen as a land 'uze' matter), to setting material considerations in respect of the location, siting and layout of schemes, plus specific requirements on the inclusion of women-related policies in all development plan documents (allowing for local conditions and variations, of course). Such measures would tie in with the trend towards quality assurance and the assessment of performance in development control. To complement this, EO codes (p. 172) are also needed for women in planning to create solid performance criteria against which employers' actions can be measured, covering staff recruitment, promotion, childcare and conditions of service. This should enable more women planners to achieve positions of seniority from which they will be able to exercise a more positive influence on urban policy for the benefit of all women in society. Even though I dread opening the door to the lawyers, the only way to span the dichotomy (p.16) between words and deeds, is to have mandatory laws, as men have had a time of grace to put their house in order and have not taken it. Greater vision is also needed in planning education and practice (Moser, 1993). Currently, the ability to demonstrate *man*agerial ability is essential for progression: a gendered attribute, which has little to do with whether a person has enlightened ideas and constructive policies about the built environment. Control of space no longer appears to dominate, because it is masked by a veil of managerialism. Women have remarked upon a creeping managerialism – anti-academic and anti-practical in nature – which is taking over much of the world of planning. Again, what is needed is not more waffly EO statements but the setting of precise objectives and parameters, backed up by formal monitoring, quality assurance measures and disciplinary measures for code infringement. At present, many women consider men do not take women and planning seriously, remarking of senior management, 'they do not seem to think it applies to them; but is just for young women'.

Other possibilities are emerging. Some women planners are questioning what has really been gained by their becoming planning officers, and entering the world of men. In spite of the current property recession, some are reflecting upon what might be achieved in operating from within the private-sector world of property development, not as powerful incipient capitalists, but as deviants, as outcast traders seeking alternative paths of influence over the built environment. Some are investigating the mileage in voluntary organizations such as housing associations, planning aid groups and national networks and data bases. Others are being drawn to ancient, private-realm forms of zone zapping power, evaluating the potential of spiritual forces, and the scope of their own nurturing powers over the next generation as a means of shaping the future. But, most of all, many are saying: to make progress in the built environment professions changes are needed in the domestic sphere of home, family and marriage, with a reconceptualization of roles and responsibilities between men and women, and attitudes and assumptions in general. Much of our zone zapping has been one way. Women have penetrated the public realm, but there has not been a commensurate change in man's role and responsibilities in the private realm. The public/private dichotomy itself has not been broken down but is still full of power. Domestic labour must be reallocated more fairly between men and women, or reconceptualized as a public responsibility. The issue of childcare, including synchronization of school hours with working times, has scarcely begun to be addressed. An enormous gap exists between the illusion of equal employment opportunities for all, and the reality of inadequate childcare provision for many. Tackling this private issue publicly is fundamental to implementing genuine equal opportunities policy. In conclusion, I believe the women and planning movement will continue evolving, and the form of its future development will be innovative, suprising, unexpected, risk-taking, blessed and powerful.[2]

NOTE

1 Some of the policies referred to in LWPG, 1991, from Draft Unitary Development Plans were subsequently rejected by the D.o.E. planning inspectors' at Inquiry, because they were seen as *ultra vires*, that is 'not a land use matter'. Some UDP women's policies were also seen as too detailed for a development plan, being concerned with 'quotas' (which the D.o.E. particularly disliked) on, for example, creche provision, customer toilets, baby buggy parking space, whereas no similar problem was encountered with car parking requirements. Women's policies fall foul of the general policies/detailed standards dichotomy, and also the insides/outsides debate in respect of the extent of a planners control over the design of a development internally as well as externally. The London Women and Planning Forum, in association with WDS and the Association of London Planning Authorities is monitoring the situation as there are undoubtedly disparities between individual Planning Inspectorates' decisions on such places.
2 I invite the reader to write out her/his own list of contents for such a code. In Appendix III, as an intermediate stage, I set out my own list of areas which require change, and some planning policy guidelines, as a basis for further discussion.

APPENDIX I
Women in planning

1 Membership of the professional bodies as at December 1993

Body	Full members (% female)		Student members (% female)		Total members (% female)	
RTPI	14,341	(18.1)	3,094	(42.3)	17,435	(22.5)
RICS	68,708	(5.2)	21,272	(15.2)	89,980	(7.6)
ICE	49,610	(1.0)	9,733	(10.0)	77,966	(3.5)
ISE	17,175	(1.5)	6,316	(11.3)	23,491	(4.2)
CIOB	25,118	(0.7)	9,439	(4.3)	33,557	(1.7)
CIOH	4,052	(43.0)	4,331	(56.0)	11,644	(45.0)
ASI	5,164	(1.1)	301	(4.3)	5,465	(1.2)
ISVA	5,444	(5.6)	1,913	(17.0)	7,357	(8.5)
RIBA	28,126	(7.0)	3,326	(25.0)	31,452	(7.5)
CIBSE	12,869	(1.2)	2,525	(4.9)	15,394	(1.8)
LI	2,144	(38.0)	353	(38.0)	3,695	(40.0)
NAEA	(non examining body)				9,494	(20.0)

Source: The professional bodies as at December 1993. See Greed (1993: Table 3.1) for 1991 situation. In some cases there are other intermediate or honorary categories which make up the remainder of the total, who are not strictly speaking either fully qualified members or students, such as probationers, technicians, international members, graduate associates. Also categories of data may be redefined by professional bodies since 1991 version.

Key

Female percentages in brackets.
RTPI   Royal Town Planning Institute
RICS   Royal Institution of Chartered Surveyors
ICE   Institution of Civil Engineers
ISE   Institution of Structural Engineers
CIOB   Chartered Institute of Building
IOH   Chartered Institute of Housing
ASI   Architects and Surveyors Institute
ISVA   Incorporated Society of Valuers and Auctioneers
RIBA   Royal Institute of British Architects
CIBSE   Chartered Institution of Building Services Engineers
NAEA   National Association of Estate Agents
LI   Landscape Institute

2 Women members of the Royal Town Planning Institute

Date	Total membership (student and corporate)	Corporate	Percentage of females	
			Corporate	Students
1920	200			
1930	400			
1940	1,000			
1950	2,500			
1960	4,000			
1970	8,000	4,424		
1971			5.4	9.9
1975			6.5	14.3
1978			7.2	16.3
1980	12,500	7,600		
1983			10.5	19.6
1986			12.6	28.6
1990	15,956	12,071	15.0	37.0
1991			16.5	38.5
1992	17,109	12,809	17.8	40.9

Source RTPI records (earlier figures are approximations, owing to war-time loss of files 1914–38, and non-gender specific records).
Note Between 1920 and 1960, it is estimated there was a gradual growth in numbers of women from under 1 per cent in 1920: in reality just a handful of notable individuals. *But* many women were active in the wider town planning movement outside of Institute membership.

3 Gender divisions in planning education in Britain

(a) 1991

Institution	Undergraduate		Postgraduate	
	Full-time	Part-time	Full-time	Part-time
Polytechnics	628 (m)	135 (m)	9 (m)	382 (m)
	343 (f)	96 (f)	9 (f)	328 (f)
Universities	561 (m)	0 (m)	157 (m)	142 (m)
	362 (f)	0 (f)	108 (f)	97 (f)
Total	1189 (m)	135 (m)	166 (m)	524 (m)
	705 (f)	96 (f)	117 (f)	425 (f)

(b) 1992

Institution	Undergraduate		Postgraduate	
	Full-time	Part-time	Full-time	Part-time
Polytechnics	444 (m)	n/a	16 (m)	228 (m)
	240 (f)	n/a	16 (f)	226 (f)
Universities	620 (m)	0	200 (m)	176 (m)
	396 (f)	0	147 (f)	135 (f)
Total	1064 (m)	n/a	216 (m)	464 (m)
	636 (f)	n/a	163 (f)	361 (f)

Source Royal Town Planning Institute, September 1992.
Notes In October 1992 the polytechnics were renamed universities, generally being referred to as 'new universities' as against the 'old' existing universities.

NB As to other 'minorities', under 4 per cent of students are of non-Anglo ethnic minority origin (but higher in some London colleges). In 1992 registered disabled students consisted of 2 undergraduate part-timers and 2 postgraduate part-timers only (gender not specified).

4 Women's employment in local government planning departments

Employment of planning staffs in planning departments comprised 16, 612 total in 1991, made up of 11, 279 (58 per cent) professional grade and 8,333 (42 per cent) other staff (adminstrative, clerical, etc.). Fifty-two per cent of professional staff were employed in shire districts; 20 per cent in shire counties; 18 per cent in metropolitan districts; and 11 per cent in London boroughs (for explanation of terms and organization of planning system, see Greed, 1993: 19–32). Twenty-three per cent of all professionally qualified staff in planning offices in England and Wales were female. Only 2 per cent of posts available were part-time, and 83 per cent of these posts were taken by women. Women comprised 65 per cent of non-professional staff in planning offices. Women comprised 2 per cent of chief planning officers and 5 per cent of deputy chief planning officers, but 32 per cent of 'officer' (ordinary/regular) grade, and 41 per cent of junior officer grade. London boroughs had the highest number of females employed in full-time professional posts (29 per cent). Childcare benefits (above statutory requirements) such as crèche provision and enhanced maternity and childcare benefits were offered by approximately 5 per cent of employers; paternity leave was available in 15 per cent of offices; and career break schemes in approximately 6.5 per cent of offices. Turnover of full-time professional staff per annum was 11.6 per cent for females, and 8.2. per cent for males; part-time turnover was 6.3 per cent for females and 11.1 per cent for males.

Source Summary of Local Government Management Board: Survey of planning staffs in local authority planning departments as at 31 October 1991 (LGMB, 1992), circulated by RTPI. This provides a dazzling array of material on all aspects of planning employment, although information on ethnicity and disability does not appear conspicuously. Note, 20 per cent of planners are employed in the private sector and are subject to other conditions of service (some worse, some better, cf. RICS, 1990).

5 RTPI membership as at December 1993

Category	Total	Female	Non-white*	Disabled*
Fellows	613	20	15	5
Members	12,688	2,553	364	35
Students	3,094	1,309	180	4
Legal	144	8		
Honorary	63	6		
International	26	8	4	
Retired	807	28	8	28
All	17,435	3,932	571	72

Source: RTPI.
* Non-white and disabled not divided into m/f, nor ethnic grouping.

Summary
22 per cent Female, 3.3 per cent 'non-white, 0.4 per cent disabled.
N.B. 80 per cent of town planners are employed in government bodies, mainly local government, but 80 per cent of surveyors (and generally other built environment professions) are to be found in the private sector.

APPENDIX II
Key texts on women and built environment and planning

The purpose of this section is to give key references (and signpost themes) in studying 'women and planning'.

FOUNDATIONAL TEXTS

Texts presenting principles of feminist theory, and thus proving the existence of patriarchy, include: Spender, 1983; Mitchell and Oakley, 1986, McDowell and Pringle, 1992.

Feminist research: Roberts, 1981; Stanley and Wise, 1990, 1993.

Women's position in society: de Beauvoir, 1949; Bernard, 1981; Whitelegg et al., 1982.

'Famous' second-wave texts include: Friedan, 1963; Greer, 1973; Millett, 1970; Daly, 1973.

HISTORY OF PLANNING

Malestream texts include: Bacon, 1967; Bor, 1972; Burke, 1971; Burke and Taylor, 1990; Cherry, 1981; Doxiadis, 1968; Dyos, 1968; Morris, 1972; Mumford, 1965; Summerson, 1978; RTPI, 1986; and see Greed, 1993a, chapter 8).

Feminist critiques related to the mainstream account of 'the history of planning' include: Morgan, 1974 and 1978; Greer, 1979; Spender, 1982; LFHG, 1983; Levine, 1990; Miles, 1988; Goodison, 1990; Wilson, 1991; Boulding, 1992, Vols I and II.

Work relevant to the history of 'women and planning' includes: Morgan, 1974 (one of the first British, second-wave, urban feminist books); Darley, 1990; Ravetz, 1980; Hayden, 1981; Bernard, 1981; Boyd, 1982; Pearson, 1988.

Material on utopias includes: Kanter, 1972; Hayden, 1976; Darley, 1978; Goodwin, 1978; Manuel, 1979; Taylor, 1984; Albinski, 1988; Bammer, 1991.

Historical texts include: de Pizan, 1405 (earliest known urban feminist book); Gilman, 1915, 1921 (see chapter 6 for first-wave urban feminism).

The following are key texts in the development of second-wave feminism itself: de Beauvoir, 1949; Friedan, 1963; Millett, 1970; Greer, 1973, and Whitelegg et al., 1982; whilst Jacobs, 1964 prefigures second wave urban feminism in all but name.

WOMEN AND PLANNING

In the early 1980s a series of encyclopaedic books from North America dealing with a wide range of urban feminist issues past and present was published, which influenced the development of the British women and planning movement. These include: Torre, 1977; Wekerle et al., 1980; Keller, 1981; Stimpson et al., 1981; Hayden, 1981, 1984.

The movement was also influenced by other international urban feminist research (IJURR, 1978), especially by the work of women geographers, the ideas of British feminist geography being encapsulated in the writings of the Women and Geography study group (WGSG, 1984), and by Little et al. (1988). A still-flourishing Canadian periodical *Women and Environments* (published in Toronto) established in the 1970s has acted as a transmitter. Several of the mainstream urban journals have devoted a 'special' issue to the topic (*Built Environment*, 1984; *Ekistics*, 1985; *TCPA*, 1987). Also *WEB* (Women and the Built Environment) produced (approximately quarterly) by Women's Design Service, London, covers topical issues, whilst WEN (women's environmental group) and WIP (women in property) and GASAT (gender and science and technology) occasionally publish.

The following are broader texts on women and space: Ardener, 1981; Attfield and Kirkham, 1989; Cockburn, 1977, 1983, 1985; McDowell, 1983; Squier, 1984; Delphy, 1984; Massey, 1984; Mackenzie, 1989; Greed 1991; Roberts, 1991; Wajcman, 1991; Wilson, 1991.

Books which link class and gender include: Davis, 1981; Crompton and Jones, 1984; Arber, et al. 1986; Acker, 1988; Delamont, 1989; Crompton and Mann, 1986; Crompton and Sanderson, 1990. Spain, 1992 and Colomina, 1992 on sexuality and space.

WOMEN AND PLANNING POLICY

By the mid 1980s 'women and planning' policies were gradually being included in local planning authority documents. London took the lead at first, see all GLC references, and publications of the London Boroughs Women and Planning Group (LWPG, 1991), and planning aid groups (LPAS, 1986a, 1986b)). Over the last ten years a range of cities have produced material, such as Sheffield, Birmingham, Southampton (q.v.), and Scotland (Brand, 1985; Chalmers, 1986; CSLA, 1991) and Ireland (Barrett et al., 1991).

Dissertations at various degree levels (including Bhride, 1987; Barnes, 1989; Taylor, J., 1990; Vaiou, 1990) have been produced on aspects of women and planning.

The question of planning for women is discussed in a range of texts including, Morphet, 1983; Taylor, 1985, 1988; Taylor, N., 1986; Morris, 1986; Poulton and Hunt, 1986; Howatt, 1987; TCPA, 1987; RTPI, 1987, 1989b; Bulos, 1987; Taylor, J., 1990; Foulsham, 1990; Sandercock and Forsyth, 1992; Greed, 1993a chapter 12; Little, 1994; Maser 1993.

The following references are particularly relevant to the named topics.

Childcare Leach, 1979; Caffrey and Eanaigh, 1990; Greenwich, 1991; and check GLC.

Design Anscombe, 1984; Attfield and Kirkham, 1989; McQuiston, 1989; and check WDS.

Disability LBDRT, 1991, CSLA, 1991, and many access reports; and GLC.

Employment Schreiner, 1911; Brueghal, 1983; Mc.Dowell, 1991.

Environment (and countryside) Carson, 1965; Caldecott and Leland, 1983; Griffin, 1984; Shoard, 1980, 1987; Little, 1990; Whatmore, 1991.

Housing Brion and Tinker, 1980; MATRIX, 1984; Coleman, 1985; Leevers, 1986; Levinson and Atkins, 1987; Roberts, 1991; Sprague, 1991.

Race Davis, 1981; RTPI, 1983; CCCS, 1992; Amoo-Gottfried, 1988; Ahmed 1989; Griffith and Amooquaye, 1989; Smith, 1989; Krishnayaran, 1990; Benzerfa-Guerroudj, 1992; GLC; Cross and Keith, 1993.

Retail Pain, 1967; Sheffield, 1986; Mackeith, 1986; Bowlby, 1988; Birmingham for people, 1989; Scott, 1989.

Safety Manchester, 1987; Lamplugh, 1988; Southampton, 1991; Birmingham City Council, 1991; check GLC.

Sport and Leisure Deem, 1986.

Third World Rogers, 1980; Momsen and Townsend, 1987; Moser, 1993; Moser and Peake, 1987; Rakodi, 1991.

Technology Kirkup and Keller, 1992; Wajcman, 1991.

Transport Pickup, 1984; Bhride, 1987; Grieco, 1989; Birmingham City Council, 1991; Hass-Klau *et al.*, 1992.

WOMEN IN PLANNING

Useful texts on women's position as town planners, and the nature of the planning profession include: Riley and Bailey, 1983; Brand, 1985; Nadin and Jones, 1990; Buckingham-Hatfield, 1991; RTPI, 1987, 1989b; Howe, 1980; Howe and Kaufman, 1981; Estler *et al.*, 1985 demonstrate how planners' normative ethics reflect their personal views and thus affect approaches to policy making.

Texts on women and local government (where most women planners work), and the wider political setting include: Stacey and Price, 1981; Webster, 1983; Rogers, 1983, 1988; Button, 1984; Siltanan and Stanworth, 1984; Halford, 1987; Barron *et al.*, 1988; Watson, 1990.

Texts relevant to women planners' experience in higher education include: Culley and Portuges, 1985; Gibbs, 1987; Rodriguez-Bachiller, 1988 on the nature of the subjects taught; Joseph, 1978 and Greed, 1991 on surveying education; Bunch and Pollock, 1983 on how they are taught; Weisman and Birkby, 1983, gives women's alternatives to planning education. Langland and Gove, 1981, and Acker, 1984 consider the wider academic setting. Whyte *et al.*, 1985, and Weiner, 1985 reflect on girls' schooling. WDS, 1993 explains the dynamics at work in architectural education, warning that after a peak in the late 1980s the number of women entering architecture courses has dropped slightly, and women's course completion rate is still declining. In comparison the Women and Housing Working Party Report (Levison and Atkins, 1987) demonstrates women's lack of progress in spite of the apparently 'girl-friendly' (Whyte *et al.*, 1985) ethos of housing.

Texts on other professions: Silverstone and Ward, 1980, especially on the Equal Opportunities aspects; Collinson *et al.*, 1990; Merchant, 1993. For comparison also see:

Architecture Wigfall, 1980; MATRIX, 1984; Estler, 1985; Adam, 1987; Walker, 1989; Lorenz, 1990.

Civil Engineering Swords-Isherwood, 1985; Gale, 1989a and b; Carter and Kirkup, 1990; Kirkup and Keller, 1992; MccGwire, 1992: 89–91.

Law Holcombe, 1983; Atkins and Hoggett, 1984; Smart, 1984; Spencer and

Podmore, 1987; Law Society, 1988. A study of ethnic minority lawyers demonstrated immediate post qualification problems of not getting on the right track, in the right office, could mar a person's progress for life (King *et al.*, 1990).

Management Kanter, 1977; Hertz, 1986; UKIPG, 1990; Marshall, 1984.

Medicine Elston, 1980; Lorber, 1984, Savage, 1986.

Surveying Greed, 1991; Joseph, 1978; Thompson, 1968.

APPENDIX III
Women and planning: Policy proposals and initiatives

SUMMARY OF TOPICS FOR WHICH CHANGE IS REQUIRED: IDEAL AIMS

Private realm

Reconceptualization of roles and duties in the home for both men and women, especially in respect of childcare and housework.

Changes in all personal financial structures, including taxation, mortgages, and pensions, to the benefit of women.

Changes in attitudes, perceptions and values affecting woman's place.

Public sphere

State harmonization of all school, college and workplace hours, holidays and annual cycles.

24-hour childcare with range of facilities and choices.

Restoration of public restaurants, laundries and domestic support.

Neighbourhood concierge system, with (real) community care and support services, for elderly, disabled and children, and (as required) temporary care for pets, plants, taking in post and parcels, and home-sitting.

Professional realm

Areas which should be the subject of both policy guidance and enforceable anti-discrimination codes, and resource input:

- Awareness training, with EO as integral part of CPD.
- EO audits, both quantitative and qualitative.
- Reconceptualization of appraisal, recruitment and promotion priorities, and expected attributes, qualities and values.
- Positive discrimination policies.
- Flexible working, maternity and paternity leave, career breaks, job share, returners' programmes, retirement, superannuation.
- Networking, mentoring, caring support systems.
- Anti-sexual-harassment codes.
- Site safety and security.
- Research publication and funding.
- Careers advice and schools liaison.

WOMEN AND PLANNING GUIDELINES:
REAL-WORLD PRAGMATIC POLICIES

These guidelines relate to the world as it is, and are intended to assist local authority planners, including pragmatic means of putting theory into practice, thus moving cities gradually towards the ideals expressed in chapter 11. I produced these in 1993, originally for the consultations stage on a proposed RTPI PAN (Planning Advice Note) produced by Yorkshire women planners.

In developing policy the following principles should be applied:

Survey data should be collected to reflect fully the composition and diversity of the population in question, and should be analysed with due consideration of the respective contribution, lifestyles and needs of the various subgroups of which it is composed, especially women.

Attention should be given in public participation exercises to the timing, location, organization and approach adopted in order to reach as wide an audience as is possible, and, for example, this may involve running women-only sessions with childcare provision.

In the spirit of good practice and quality assurance all planning authorities should ensure that women planners are found at all senior staff levels, in order to ensure a full gender-related input at all stages of the plan-making process. In rural areas and authorities with low female planning officer numbers, it is recommended that a consortium is formed with adjacent planning authorities and that a panel of such women planners is identified who might be co-opted on to relevant policy-making committees as and when required by each local authority in question, as an interim measure.

In producing policy statements at development plan level for whole cities attention should be given to the social reality of women combining paid work with their childcare and home-making responsibilities. The policy implication of this might be the need to ensure that residential areas and employment locations are either in close proximity or within reasonable travelling distance, and that public transport routes accommodate the work–home journeys generated.

In producing policy statements at local plan level, consideration should be given to the location and accessibility of those land uses, facilities, and types of development most used by women. Shopping provision (especially food shopping) should be encouraged in a balanced manner within the local area. Retail outlet location should show a relationship either with residential or employment catchment populations, and local provision be given priority over applications for out-of town-centres. Adequate pedestrian and public-transport access, as well as safe shopper parking should be stipulated. In determining applications the objective should be to achieve as short distances as possible between different land uses and amenities. This will result in a multiplication, and higher intensity of local facilities, more opportunties for local employment, and eventually a multi-nuclei city form, high on 'sustain-ability'. Full integration and mixing of land uses should be encouraged where circumstances permit. Any provision of public open space should take into account the varied needs within the community on the basis of age and gender, with an emphasis on providing for women's needs to correct the current imbalance. Development of small parks, and incidental landscaped areas, and children's play areas by the private sector as an element in a planning gain agreement should be welcomed, with attention being given to long-term funding, as well as provision.

In seeking to control development at estate and individual site level, attention should be given to ensuring a fully accessible, safe environment, by means of adequate lighting and good visibility around footpaths. Blind corners, shrub planting near to footpaths, and high walls along footpaths should be avoided. In estate development, shopping malls, entrances to health and leisure facilities and other amenities open to the public, steps should only be used where no other alternative, such as ramps, is possible. All entrances to buildings and passageways should be of a width which will accommodate a double buggy (and thus accommodate other groups such as wheelchair users and laden shoppers). In all residential development, it is considered good practice, and within the ambit of planning law, for planners to have a degree of control on the internal layout and sizes of rooms, in so far as they are responsible for how 'development' is used.

In all centres of human activity, including local shopping centres, especially those which are also used as transport interchanges, public conveniences and baby changing areas should be provided, with the provision reflecting the gender breakdown of users (80 per cent of shoppers are women) (For example, 2 cubicles per 3,000 square metres of retail floor space for women, plus male facilities as appropriate to user–sex ratio). Traditional facilities, free and without turnstiles, (not superloos) should be located at ground level without steps, should be adequately lit, openly located without tree cover, and be supervised and open between 7.30 a.m. and 11.00 p.m. This may be achieved through planning gain contribution, but other means should be explored such as provision and supervision by local commercial outlets, including existing shops, cafés, and petrol stations, provided facilities are open at all times to non-customers. Non-compliance by local authorities would be taken into account as a negative factor in determining the level of central government financial allocation to the local authority in question.

In determining applications which are likely to generate jobs attention should be given, as an integral part of any economic strategy, to the gender structure of the workforce and other users, amd every opportunity should be taken to facilitate childcare provision either as part of a planning agreement under S.106 or by other statutory, or commercial means. It is recommended that 1 childcare space is provided for every 500–1,000 square feet of net office space, depending upon local circumstances, with commercial rates allowances (per square feet allocated) for such space provided surplus to the current needs of the employer. Just as car parking standards are currently restricted or relaxed to shape development patterns and office density, crèche space might be treated in a similar manner in preliminary development negotiations.

BIBLIOGRAPHY

Abbott, Pamela and Wallace, Claire (1990) *An Introduction to Sociology: Feminist Perspectives*, London: Routledge.

Abercrombie, Patrick (1943) *Town and Country Planning*, London: Oxford University Press.

—— (1945) *Greater London Development Plan*, London: HMSO.

Abrams, M. (1953) *The Mirror and the Lamp: Romantic theory and critical tradition*, Oxford: Oxford University Press.

Abrams, Philip (1992) (advertisement/press notice), *Journal of Historical Sociology*, Oxford: Blackwells (May).

Acker, Joan (1973) 'Women and social stratification. A case of intellectual sexism', *American Journal of Sociology* 78 (4): 936–45.

—— (1988) 'Class, gender, and the relations of distribution', *Signs* 13 (3): 473–97.

Acker, Sandra (1984) 'Women in higher education: what is the problem?', in Sandra Acker, and David Warren Piper, (eds) *Is Higher Education Fair to Women?* Slough: NFER-Nelson.

Adam, Peter (1987) *Eileen Gray: Architect Designer*, London: Thames & Hudson.

Adams, Carol (1990) *The Sexual Politics of Meat*, Cambridge: Polity.

Adburgham, Alison (1964) *Shops and Shopping, 1800–1914*, London: Barrie & Jenkins, 1989.

Adshead, S. (1923) *Town Planning and Town Development*, London: Methuen.

Ahmed, Yunus (1989) 'Planning and racial equality', *The Planner* 75 (32): 18–20, 1 December.

Aisbett, Alan (1990) 'Public/Private-Sector Joint Ventures: Local Government and Housing Act 1989 effects', *Estates Gazette*, pp. 24–9, 62, 24 February.

Albinski, Nan Bowman (1988) *Women's Utopias in British and American Fiction*, London: Routledge.

Alcott, Louisa May (1868) *Little Women*, London: Collins, 1958.

Aldous, Tony (1972) *Battle for the Environment*, London: Fontana.

Aldridge, Meryl (1979) *The British New Towns*, London: Routledge & Kegan Paul.

Allin, Paul and Hunt, Audrey (1982) 'Women in official statistics', in Elizabeth Whitelegg, (ed.) *The Changing Experience of Women*, London: Basil Blackwell in association with the Open University.

Alonso, William (1965) 'Cities and city planners', in Kenneth Lynn and 'Daedalus', *The Professions in America*, Boston: Houghton Mifflin.

Ambrose, Peter and Colenutt, Barry (1979) *The Property Machine*, Harmondsworth: Penguin.

Ambrose, Peter (1986) *Whatever Happened to Planning?*, London: Methuen.

BIBLIOGRAPHY

Amoo-Gottfried, Hilda (1988) 'Racism within the legal profession', *Law Society's Gazette* 85 (1): 10–20.

Anker, Stephen, Kaiserman, David and Shepley, Chris (1979) *The Grotton Papers*, London: The Royal Town Planning Institute.

Anscombe, Isabelle (1984) *A Woman's Touch: Women in design from 1860 to the present day*, London: Virago.

Arber, Sara, Dale, Angela and Gilbert, Nigel (1986) 'The limitations of existing social class classifications for women', in A. Jaccoby (ed.) *The Measurement of Social Class*, Department of Sociology, University of Surrey.

Ardener, Shirley (ed.) (1978) *Defining Females: the nature of women in society*, London: Croom Helm.

—— (ed.) (1981) *Women and Space: Ground rules and social maps*, London: Croom Helm.

Armitt, Lucie (1991) *Where No Man has Gone Before: Women and science fiction*, London: Routledge.

Armstrong, David (1993) 'Public health spaces and the fabrication of identity', *Sociology* 27, (3): 393–410.

Arthur, Chris (1991) 'Diversion around the God slot: how can the humanities ignore the human soul?', *The Times Higher Education Supplement*, 8 November.

Ashworth, William (1968) *The Genesis of British Town Planning*, London: Routledge & Kegan Paul.

Atkins, Susan and Hoggett, Brenda (1984) *Women and the Law*, Oxford: Blackwell.

Attar, Dena (1990) *Wasting Girls' Time: The history and politics of home economics*, London: Virago.

Attfield, Judy (1989) 'Inside Pram Town, a case study of Harlow house interiors, 1951–61', in Judy Attfield and Pat Kirkham (eds) *A View from the Interior: Feminism, women and design*, London: Women's Press.

Attfield, Judy and Kirkham, Pat (eds) (1989) *A View from the Interior: feminism, women and design*, London: Women's Press.

Attwood, Lynne (1990) *The New Soviet Man and Woman: Sex-role socialisation in the USSR*, London: Macmillan in association with the Centre for Russian and East European Studies, University of Birmingham.

Atwood, Margaret (1987) *The Handmaiden's Tale*, London: Virago.

Augustine (AD 4–26) *City of God (De civitate dei)*, London: Heinemann, 1967.

Babbit, I. (1947) *Rousseau and Romanticism*, Cleveland: World Publicity Company, Ohio.

Bacon, Edmund (1967) *Design of Cities*, London: Thames & Hudson.

Bailey, Joe (1975) *Social Theory for Planning*, London: Routledge & Kegan Paul.

Balchin, Paul and Bull, Gregory (1987) *Regional and Urban Economics*, London: Harper & Row.

Ball, M. (ed.) (1985) *Land Rent, Housing and Urban Planning: A European perspective*, London: Croom Helm.

Ball, Michael (1988) *Rebuilding Construction: Economic change in the British construction industry*, London: Routledge.

Bammer, Angelika (1991) *Partial Visions: Feminism and utopianism in the 1970s*, London: Routledge.

Banks, Olive (1981) *Faces of Feminism: A study of feminism as a social movement*, Oxford: Martin Robertson.

Barclay, Irene (1980) 'Recollections of a pioneer', *Chartered Surveyor* 113 (4): 246–8, November.

Barker-Benfield, Ben (1972) 'The spermatic economy: a nineteenth century view of sexuality', *Feminist Studies* 1 (1): 45–74.

Barlow Report (1940) *Report of the Royal Commission on the Distribution of the Industrial Population*, London: HMSO, Cmd. 6153.

Barnes, Deborah (1989) 'Women and planning: Recent initiatives with reference to the GLC and Sheffield City Council', unpublished BA (Hons) dissertation, Oxford Polytechnic, Department of Town Planning.

Barrett, Maeve, Boothman, Geraldine, Caffrey, Claire, McCambley, Sheena, McElwee, Dympna, McKeown, Gabrielle, Eanigh, Cristín Ní O'Shee, Majorie and Sheridan, Anne-Marie (1991) *Submission from the Women and Planning Group for the Review of Dublin County Council Development Plan*, Dublin: Women and Planning Group, Irish Planning Institute, and RTPI, Irish Branch, Southern Section.

Barron, Jacqueline, Crawley, Gerald and Wood, Tony (1988) *Married to the Council?: The private costs of public service*, Bristol: Bristol Polytechnic, Department of Economics and Social Science.

Barthes, Roland (1973) *Mythologies*, London: Paladin.

Bassett, Keith and Short, John (1980) *Housing and Residential Structure, Alternative Approaches*, London: Routlege & Kegan Paul.

Battersby, Christine (1989) *Gender and Genius: Towards a feminist aesthetics*, London: The Women's Press.

Beauchamp, Diana and Hudson, Roger (1985) *Our First Fifty Years: 1935–1985: Women's Gas Federation and Young Homemakers*, London: Women's Gas Federation and Young Homemakers.

Beauvoir, Simone de (1949) *The Second Sex*, New York: Vintage, 1974.

Bebbington, David (1979) *Patterns in History*, London: Inter-Varsity Press.

Beckett, Jane and Cherry, Deborah (1987) *The Edwardian Era*, London: Phaidon Press and Barbican Art Gallery.

Bell, Colin and Bell, Rose (1972) *City Fathers: The early history of town planning in Britain*, Harmondsworth: Penguin.

Bell, Florence (1911) *At the Works*, London: Thomas Nelson.

Bellamy, Edward (1982, 1888) *Looking Backward: 2000–1887*, Harmondsworth: Penguin.

Bellos, Linda (1988) 'Things move on', in Amanda Sebestyen, *'68, '78, '88: From women's liberation to feminism*, Bridport, Dorset: Prism Press.

Benevelo, Leonardo (1976) *The Origins of Modern Town Planning*, London: Routledge & Kegan Paul.

Benzerfa-Guerroudj, Zina (1992) 'Les femmes algériennes dans l'espace public', in *Architecture et Comportement*, 8 (2): 123–36, Lausanne: Association de la revue 'architecture et comportement'.

Berger, Peter and Luckman, Thomas (1972) *The Social Construction of Reality*, Harmondsworth: Penguin.

Bernard, Jessie (1981) *The Female World*, New York: Free Press.

Bernstein, Basil (1975) *Class, Codes, and Control*, London: Routledge.

Bhride, Karla Ni (1987) 'Women and planning: an analysis of women's mobility and accessibility to facilities', Dublin Thesis 1987, Master of Regional and Urban Planning Degree, No. 182, unpublished thesis.

Birmingham City Council (1988) *Women and Transport*, Birmingham: Transportation Policy Division.

—— (1991) *Women Safety and Design in City Centres*, Birmingham: Department of Planning and Architecture.

Birmingham for People (1989) *Women in the Centre: women, planning and Birmingham city centre*, Birmingham: Birmingham for People, Women's Group.

Blowers, Andrew (1993) (ed.) *Planning for a Sustainable Environment: Report of the Town and Country Planning Association*, London: Town and Country Planning Association, in association with Earthscan.

Blomfield, Margaret (1943) *Our Towns: A close up*, Oxford: Oxford University Press.

Blumer, Herbert (1965) 'Sociological implications of the thought of George Herbert Mead' in B. Cosin, I. Dale, G. Esland, D. MacKinnon and D. Swift (1977) *School and Society: A sociological reader*, London: Routledge, & Kegan Paul.

Boardman, Philip (1978) *The World of Patrick Geddes*, London: Routledge.

Bologh, Roslyn (1990) *Love or Greatness: Max Weber and masculine thinking, a feminist inquiry*, London: Unwin Hyman.

Bolsterli, Margaret (1977) *The Early Community at Bedford Park: The pursuit of corporate happiness in the first garden suburb*, London: Routledge & Kegan Paul.

Bookchin, Murray (1992) *Urbanisation without Cities: The rise and decline of citizenship*, Montreal: Black Rose Books.

Booth, Charles (1889–1906) *Life and Labour of the People of London*, Chicago: University of Chicago Press, 1968.

Booth, Philip (1986) 'The teaching of planning history in Britain', *Planning History Bulletin* 8 (1): 21–7.

Booth, William (1890) *In Darkest England and the Way Out*, London: Salvation Army.

Bor, Walter (1972) *The Making of Cities*, London: Leonard Hill.

Boulding, Elise (1992) *The Underside of History*, 2 vols, London: Sage.

Bowlby, Sophie (1988) 'From corner shop to hypermarket: women and food retailing', in Jo Little, Linda Peake, and Pat Richardson, (eds) *Women and Cities, Gender and the Urban Environment*, London: Macmillan.

—— (1989) 'Gender issues and retail geography', in Sarah Whatmore, and Jo Little (eds) *Geography and Gender*, London: Association for Curriculum Development in Geography.

Boyd, Nancy (1982) *Josephine Butler, Octavia Hill, Florence Nightingale: Three Victorian women who changed the world*, London: Macmillan.

Brand, Janet (1985) *Women and Planning in Scotland: Report of conference*, Strathclyde: Department of Urban and Regional Planning, University of Strathclyde.

Brandon, Ruth (1991) *The New Women and the Old Men: Loves, sex and the woman question*, London: Harper-Collins, Flamingo.

Brazier, Chris (1991) 'A journey to the heart of Vietnam', *New Internationalist* 216: 4–6, February.

Brett, Meta (ed.) (1989, annual) *Directory of Official Architecture and Planning, 1990*, London: Longman.

Briggs, Asa (1968) *Victorian Cities*, Harmondsworth: Penguin.

Brion, Marion and Tinker, Anthea (1980) *Women in Housing: Access and influence*, London: Housing Centre Trust.

Broady, Maurice (1968) *Planning for People*, London: NCSS/Bedford Square Press.

Broude, Norma and Garrard, Mary (1982) *Feminism and Art History: Questioning the litany*, New York: Harper & Rowe.

Brownmiller, Susan (1975) *Against our Will: Men, women and rape*, London: Secker & Warburg.

Brueghal, Irene (1983) 'Women's employment legislation and the labour market', in Jane Lewis (ed.) *Women's Welfare, Women's Rights*, London: Croom Helm.

Bruton, Michael (1975) *Introduction to Transportation Planning*, London: Hutchinson.

Bryson, Valerie (1992) *Feminist Political Theory: An introduction*, London: Macmillan.

Buchanan, Colin (1958) *Mixed Blessing: The motor in Britain*, London: Leonard Hill.

—— (1963) *Traffic in Towns*, Harmondsworth: Penguin.

—— (1972) *The State of Britain*, London: Faber.

Buckingham-Hatfield, Sue (1991) 'Equal opportunities survey', in *Newsletter*, Spring 1991: 3, Institute of British Geographers, Urban Geography Group.

Buckley, Mary (1989) *Women and the Ideology of the Soviet Union*, Hemel Hempstead: Harvester Wheatsheaf.

Built environment (1984) Special Issue on 'Women and the built environment', 10 (1).

Bulmer, Martin (1984) *The Chicago School of Sociology*, London: University of Chicago Press.

Bulos, Marjorie (ed.) (1987) *Women and the Built Environment: An annotated bibliography to promote curriculum development in higher and further education*, London: Department of Planning, Housing and Development, South Bank Polytechnic.

—— (ed.) (1990) *Making a Place for Women: A resource handbook on women and the built environment*, London: Department of Planning, Housing and Development, South Bank Polytechnic.

Bunch, Charlotte and Pollock, Sandra (eds) (1983) *Learning our Way*, New York: Crossing Press.

Bunyan, John (1678) *Pilgrim's Progress: from this world to that which is to come*, London: Collins, 1958.

Burke, Gerald (1971) *Towns in the Making*, London: Arnold.

—— and Taylor, Tony (1990) *Town Planning and the Surveyor*, Reading: College of Estate Management.

Bushnell, G. (1968) *The First Americans: the pre-Columbian civilisations*, London: Thames & Hudson.

Button, Sheila (1984) 'Women's committees: a study of gender and local government policy formulation', *SAUS Working Paper 45*, Bristol: School of Advanced Urban Studies, University of Bristol.

Cadbury, George, Jr (1915) *Town Planning*, London: Longman.

Caffrey, Claire and Eanaigh, Ní, Cristín (1990) *Toilets, Babycare and Creche Facilities in Shopping Developments*, Dublin: Women and Planning Group.

Caldecott, Leonie and Leland, Stephanie (1983) *Reclaim the Earth: Women speak out for life on earth*, London: Women's Press.

Calder, Mick, Cavanagh, Steve, Eckslin, Claire, Palmer, Jackie and Stell, Andrew (1993) *Women and Development Plans*, Working Paper 27, Newcastle upon Tyne: University of Newcastle, Department of Town and Country Planning.

Callaway, Helen (1987) *Gender, Culture and Empire*, Urbana: University of Illinois Press.

Campbell, Beatrix (1985) *Wigan Pier Revisited: Poverty and politics in the eighties*, London: Virago.

Cardinal, Marie (1991) *Devotion and Disorder*, London: Women's Press.

Carey, Lynette and Mapes, Roy (1972) *The Sociology of Planning: A study of social activity on new housing estates*, London: Batsford.

Carr, E. (1965) *What is History?*, Harmondsworth: Penguin.

Carson, Rachel (1965) *Silent Spring*, Harmondsworth: Penguin.

Carter, Ruth (1991) *Women of the Dust*, Bristol Old Vic (play).

Carter, Ruth and Kirkup, Gill (1990) *Women in Engineering*, London: Macmillan.

Casson, Hugh (1978) *Spirit of the Age*, London: BBC Publications.

Castells, Manuel (1977) *The Urban Question*, London: Arnold.

CCCS (1992) *The Empire Strikes Back*, London: Routledge, with the Centre for Contemporary Cultural Studies, Birmingham University.

Chadwick, Edwin (1842) *Report on the Sanitary Condition of the Labouring Population of Great Britain: and on the means of its improvement*, London.

Chadwick, Whitney (1990) *Women, Art and Society*, London: Thames & Hudson.

Chalmers, Marion (ed.) (1986) *Women and Planning in Scotland: new communities –*

did they get it right?, Women and Planning Working Party, Royal Town Planning Institute, Scottish Branch, and West Lothian District Council.

Chapin, Francis Stuart, (1965; and 1979 edition with J. Kaiser) *Urban Land Use Planning*, Illinois: University of Illinois Press, 12–25.

Cherry, Gordon (1974) *The Evolution of British Town Planning*, London: Royal Town Planning Institute Edition.

—— (ed.) (1981) *Pioneers in British Town Planning*, London: Architectural Press.

—— (1984) 'The Institute 1914–1984', *The Planner* 70 (9): 8–10 September.

—— (1988) *Cities and Plans: The shaping of urban Britain in the nineteenth and twentieth centuries*, London: Edward Arnold.

—— (1991) 'Planning history: recent developments in Britain', *Planning Perspectives* 6: 33–45.

Christaller, Walther (1933) *Die Zentralen Orte Süddeutschlands*, Jena.

CILT (1990) *Survey of British Rail and Underground Stations in Hackney*, London: Centre for Independent Transport Research.

CISC (1992a) *Occupational Standards for Technical, Managerial and Professional Roles in the Construction Industry*, London: Construction Industry Standing Conference.

CISC (1992b) *Competence, Beyond the Technical: Components of creativity*, London: Construction Industry Standing Conference.

Clark, Burton (1983) 'Belief', in Burton Clark, *The Higher Education System: Academic organization in cross-national perspective*, Berkeley: University of California Press.

Clark, John (1933) *Outlines of the Law of Housing and Planning: including public health, highways, and the acquisition of land*, London: Pitman.

Clark, Kenneth (1969) *Civilisation*, London: British Broadcasting Corporation, in association with John Murray.

Clark, Martin (1991) 'Developments in human geography: niches for a Christian contribution', *Area: Institute of British Geographers* 23 (4): 339–44, December.

Clarke, Linda (1992) *Building Capitalism: Historical change and the labour process in the production of the built environment*, London: Routledge.

Clements, Barbara, Engel, Barbara, and Worobec, Christine (1991) *Russia's Women: Accommodation, resistance, transformation*, Oxford: University of California Press.

Cockburn, Cynthia (1977) *The Local State: Management of people and cities*, London: Pluto.

—— (1983) *Brothers – Male Dominance and Technological Change*, London: Pluto.

—— (1985) *Machinery of Dominance*, London: Pluto.

—— (1991) *In the Way of Women: Men's resistance to sex equality in organisations*, London: Macmillan.

Cole, G. D. H. (1945) *Building and Planning*, London: Cassell.

Coleman, Alice (1985) *Utopia on Trial*, London: Martin Shipman.

Colenutt, Bob (1992) 'Town planning in the 21st century', unpublished conference proceedings, South Bank University, London, 28 October.

Collard, Andrée and Contrucci, Joyce (1988) *Rape of the Wild: Man's violence against animals on the earth*, London: Women's Press.

Collier, Emma (1991) 'Women and public space: an exploration of gender issues in the built environment', unpublished MA dissertation. Oxford: Joint Centre for Urban Design with Oxford Polytechnic.

Collins, George and Crasemann Collins, Christiane (1965) *Camillo Sitte and the Birth of Modern City Planning*, New York: Random House.

Collins, M. P. (1989) 'A review of 75 years of planning education at University College London', *The Planner*, Special Issue, 75 (9): 23 June, pp. 18–22.

Collins, Randall (1979) *The Credential Society*, New York: Academic Press.

Collinson, D. L., Knight, D., and Collinson, M. (1990) *Managing to Discriminate*, London: Routledge.

Colomina, Beatriz (1992) *Sexuality and Space*, New York: Princeton Architectural Association.

Comfort, Alex (1979) *The Joy of Sex*, London: Quartet.

Connell, Robert (1987) *Gender and Power*, Cambridge: Polity Press.

Cooke, Philip (1987) 'Clinical inference and geographical theory', *Antipode: A Radical Journal of Geography*, 19 (1): 68–78, April.

Coole, Diana (1988) *Women in Political Theory*, London: Wheatsheaf.

Coote, Anna and Pattullo, Polly (1990) *Power and Prejudice: Women and politics*, London: Weidenfeld & Nicolson.

Corbusier, Le (1929) *The City of Tomorrow*, London: Architectural Press, (1971).

Corbusier, Le (1930) *L'Architecture d'Aujourd'hui*, Paris.

Corrin, Christine (forthcoming) 'Women in city and countryside', in Rosalind Marsh (ed.) *Women in Russia and the Former USSR*, Cambridge: Cambridge University Press, forthcoming.

CRE (Commission for Racial Equality) (1989) *A Guide for Estate Agents and Vendors*, London: CRE.

Creese, Walter (1967) *The Legacy of Raymond Unwin: A human pattern for planning*, Cambridge, Mass.: MIT.

Crompton, Rosemary and Jones, Gareth (1984) *White-collar Proletariat: Deskilling and gender in clerical work*, London: Macmillan.

Crompton, Rosemary and Mann, Michael (1986) *Gender and Stratification*, Cambridge: Polity Press.

Crompton, Rosemary and Sanderson, Kay (1990) *Gendered Jobs and Social Change*, London: Unwin Hyman.

Cross, Malcolm and Keith, Michael (eds) (1993) *Racism, the City and the State*, London: Routledge.

Crowe, Sylvia (1960) *The Landscape of Roads*, London: Unwin.

CSLA (Convention of Scottish Local Authorities) (1991) *Equal Opportunities and Planning*, Edinburgh: Convention of Scottish Local Authorities.

Culley, Margo, and Portuges, Catherine (1985) *Gendered Subjects: The dynamics of feminist teaching*, London: Routledge & Kegan Paul.

Cullingworth J. B. (1988) *Town and Country Planning in Britain*, London: Unwin Hyman.

Cunningham, Susan and Norton, Christine (1993) *Public Inconveniences: Suggestions for improvements*, London: The Continence Foundation.

Curl, James Stephens (1986) 'Review of Mackeith', *Planning Perspectives: an international journal of history, planning and the environment*, 1 (2) 187–8, May.

Dale, Jennifer and Foster, Peggy (1986) *Feminism and State Welfare*, London: Routledge.

Daly, Mary (1973) *Beyond God the Father: Towards a philosophy of women's liberation*, London: Women's Press, 1991.

—— (1993) *Outercourse: the be-dazzling voyage*, London: Women's Press.

Darke, Roy (1990) 'Introduction to popular planning', in John Montgomery and Andy Thornley, (1990) *Radical Planning Initiatives: new directions for urban planning*, Aldershot: Gower.

Darley, Gillian (1978) *Villages of Vision*, St Albans: Granada.

—— (1990) *Octavia Hill: A life*, London: Constable.

Davidoff, Leonore and Hall, Catherine (1983) 'The architecture of public and private life. English middle-class society in a provincial town 1780–1850', in D. Fraser and A. Sutcliffe (eds) *The Pursuit of Urban History*, London: Edward Arnold.

—— —— (1987) *Family Fortunes: Men and women of the English middle classes, 1780–1850*, London: Hutchinson.

Davies, Jon Gower (1972) *The Evangelistic Bureaucrat: A study of a planning exercise in Newcastle upon Tyne*, London: Tavistock.

Davies, Linda (1993) 'Aspects of equality', *The Planner*, 79 (3): 14–16, March.

Davis, Angela (1981) *Women, Race and Class*, London: Women's Press.

Davis, Mike (1990) *City of Quartz: Excavating the future in Los Angeles*, London: Vintage.

Davis, Robert (1993) *Death in the Streets: Cars and the mythology of road safety*, London: Leading Edge.

Dean, John (1990) 'Walking in the city', in Sylvia Trench and Oc Tanner, *Current Issues in Planning*, Aldershot: Gower.

Deem, Rosemary (ed.) (1984) *Co-Education Reconsidered*, Milton Keynes: Open University Press.

—— (1986) *All Work and No Play?: The sociology of women and leisure reconsidered*, Milton Keynes: Open University Press.

Deeplish (1966) *The Deeplish Study: Improvement possibilities in a district of Rochdale*, London: Ministry of Housing and Local Government.

Delamont, Sara (1976) 'The girls most likely to: cultural reproduction and Scottish elites', *Scottish Journal of Sociology* 1 (1): 29–43.

—— (1985) 'Fighting familiarity', *Strategies of Qualitative Research in Education*, Warwick: ESRC Summer School.

—— (1989) *Knowledgeable Women: Structuralism and the reproduction of elites*, London: Routledge.

—— (1992) 'Old fogies and intellectual women: an episode in academic history', *Women's History Review* 1 (1): 39–62.

Delgado, Alan (1979) *The Enormous File: A social history of the office*, London: John Murray.

Delphy, Christine (1984) (Diana Leonard trans.) *Close to Home: A materialistic analysis of women's oppresssion*, London: Hutchinson.

Denington Report (1966) *Our Older Homes: A call to action*, London: HMSO.

Department of Transport (1990) *Roads in Urban Areas*, London: HMSO.

DES (Department of Education and Science) (1966) *A Plan for Polytechnics and other colleges*, White Paper, Cmnd 3006, London: HMSO.

Descartes, René (1637) *Discourse on Method*, London: Dent, 1987.

Diamond, Derek (1986) 'Who gets what, where, why?' *Town and Country Planning* 55 (3): 98.

Dickens, Ian and Fidler, Peter (1990) *Monitoring supply: demand and employment trends in the town planning profession*, London: Royal Town Planning Institute and Birmingham Polytechnic.

Dixon, Roger and Muthesius, Stephan (1978) *Victorian Architecture*, London: Hutchinson.

D.o.E (Department of the Environment) (1971) *Sunlight and Daylight: Planning criteria and design of buildings*, London: HMSO.

—— (1972a) *How do you want to live?: A report on human habitat*, London: HMSO.

—— (1972b) *Development Plan Manual*, London: HMSO.

—— (1972c) *The Estate Outside the Dwelling: Reactions of residents to aspects of housing layout*, Design Bulletin 25, London: HMSO.

—— (1979) *Access for Disabled People to Educational Buildings*, London: HMSO.

—— (1991) *Draft Planning Policy Guidance: General policy and principles*, London: D.o.E, Development Control Policy Division, para 29 of 1.10.91 version.

—— (1992a) *Development Plans and Regional Planning Guidance: planning policy guidance, Note 12*, London: HMSO.

—— (1992b) *Development Plans: good practice guidance note*, London: HMSO.
—— (1992c) *Sport and Recreation: planning policy guidance, Note 17*, London: HMSO.
—— (1992d) *Housing and Construction Statistics*, London: HMSO.
—— (1992e) *Housing: Planning Policy Guidance Note 3*, London: HMSO.
—— (1993) *Transport: Planning Policy Guidance Note 13*, London: HMSO.
Dolan, Dennis (1979) *The British Construction Industry: An introduction*, London: Macmillan.
Dore, Ronald (1976) *The Diploma Disease*, London: Unwin.
Dorfman, Marc (1986) 'Royal architecture: how it took off', *Town and Country Planning*, 55 (2): 50–3, February.
Douglas, Mary (1966) *Purity and Danger: An analysis of the concepts of pollution and taboo*, London: Routledge & Kegan Paul, 1984.
Dower Report (1945) *National Parks in England and Wales*, London: HMSO.
Doxiadis, Constantinos (1968) *Ekistics: An introduction to the science of human settlements*, London: Hutchinson.
Drakulć, Slavenka (1987) *How We Survived Communism and even Laughed*, London: Hutchinson.
Dresser, Madge (1978) 'Review essay' of Leonore Davidoff, Jean L'Esperance, and Howard Newby (1976) *Landscape with Figures: Home and community in English society*, *International Journal of Urban and Regional Research* 2 (3) Special Issue on 'Women and the City'.
Duchen, Claire (ed.) (1992a) 'A Continent in transition: Issues for women in Europe in the 1990s', *Women's Studies International Forum* 15 (1) 1–152, Special Issue.
—— (ed.) (1992b) 'Understanding the European Community: A glossary of terms', *Women's Studies International Forum* 15 (1): 17–20.
Dudley Report (1944) *Design of Dwellings*, London: HMSO, Central Housing Advisory Committee of the Ministry of Health.
Duncan, Simon (1991) 'The geography of gender divisions of labour in Britain', *Transactions of the Institute of British Geographers* 16 (4): 420–39.
Dunleavy, Patrick (1980) *Urban Political Analysis*, London: Macmillan.
Durkheim, Emile (1948) *Elementary Forms of the Religious Life*, New York: Free Press.
—— (1897) *Suicide: A study in sociology*, London: Routledge & Kegan Paul, 1970.
Dworkin, Andrea (1983) *Right Wing Women*, London: Women's Press.
Dyos, H. (1968) *The Study of Urban History*, London: Edward Arnold.
EC (European Commission) (1990) *Green Paper on the Urban Environment*, Brussels: EEC Fourth Environmental Action Programme 1987–1992.
Ekistics (1985) *Woman and Space in Human Settlements* 52: 310, January.
Eliade, Mircea (1959) *The Sacred and the Profane: The nature of religion*, New York: Harvest, Brace and World.
Ellis, John (1991) *John Major: A personal biography*, London: Futura.
Elson, M. (1986) *Green Belts*, London: Heinemann.
Elston, Mary Anne (1980) 'Medicine', in Rosalie Silverstone and Audrey Ward (eds) *Careers of Professional Women*, London: Croom Helm.
Engels, Friedrich (1872) *Housing Question*, London: Central Books, 1975.
Enloe, Cynthia (1989) *Bananas, Beaches and Bases: Making feminist sense of international politics*, London: Pandora.
EOC (Equal Opportunities Commission) (1988) *Local Authority Equal Opportunities Policies: Report of a survey by the Equal Opportunities Commission*, Manchester: EOC.
Esher, Lionel (1983) *A Broken Wave: The rebuilding of England 1940–80*, Harmondsworth: Penguin.

Eskapa, Roy (1987) *Bizarre Sex*, London: Harper-Collins, Grafton.

Essex (1973) *Essex Design Guide*, Chelmsford: Essex County Council.

Essex, Sue (1991) 'The nature of the job', in H. Thomas and P. Healey, *Dilemmas of Planning Practice: Ethics, legitimacy, and the validation of knowledge*, Aldershot: Avebury.

Estler, Suzanne, Prussin, Labelle, Ryckman, David and Sasanoff, Robert (1985) *Gender Related Cultures and Architectural Values*, Washington: University of Washington.

Evans, Mary (1991) *Life at a Girl's Grammar School in the 1950s*, London: Women's Press.

Eve, Tristram (1948) 'Town and Country Planning Act', correspondence between Tristram Eve and Lewis Silkin, *The Journal of the Royal Institution of Chartered Surveyors* 18 (4): 204–6, October.

Faludi, Susan (1992) *Backlash: The undeclared war against women*, London: Chatto & Windus.

Filtzer, Don (forthcoming) 'Industrial working conditions and the political economy of female labour during Perestroika', in Rosalind Marsh (ed.) *Women in Russia and the former USSR*, Cambridge: Cambridge University Press, forthcoming.

Firestone, Shulamith (1979) *The Dialectic of Sex*, London: Women's Press.

Fischman, Robert (1977) *Urban Utopias in the Twentieth Century: Ebenezer Howard, Frank Lloyd Wright and Le Corbusier*, New York: Basic Books.

Fisher Mark and Owen, Ursula (eds) (1991) *Whose Cities?*, Harmondworth: Penguin.

Fitch (1985) *Shopping Centre Report, No.8, 'Women'*, London: Fitch and Company Shopping Consortium.

Fitzsimmons, Diana (1990) 'Training for planning', *The Planner* 76 (14): 29–30.

Focas, Caralampo (1989) 'A survey of women's travel needs in London', in Margaret Grieco, Laurie Pickup, and Richard Whipp *Gender, Transport and Employment*, Aldershot: Avebury.

Foley, Donald (1964) 'An approach to urban metropolitan structure', in Melvin Webber, John Dyckman, Donald Foley, Albert Guttenberg, William Wheaton, and Catherine Bauer Wurster, *Explorations into Urban Structure*, Philadelphia: University of Pennsylvania Press.

Foord, Jo and Gregson, Nicky (1986) 'Patriarchy: towards a reconceptualisation', *Antipode: a Radical Journal of Geography*, 18 (2): 186–211, September.

Fortlage, Catherine (1990) *Environmental Assessment: A practical guide*, Aldershot: Gower.

Foucault, Michel (1972) *The Archaeology of Knowledge (L'archaéologie du savoir)*, Paris: Éditions Gallimard.

—— (1976) *The History of Sexuality*, Harmondsworth: Penguin.

Foulsham, Jane (1990) 'Women's needs and planning: a critical evaluation of recent local authority practice', in John Montgomery and Andy Thornley, *Radical Planning Initiatives: New directions for urban planning in the 1990s*, Aldershot: Gower.

Frankenberg, Ronald (1970) *Communities in Britain*, Harmondsworth: Penguin.

Fraser, Nancy and Nicholson, Linda (1990) 'Social criticism without philosophy: an encounter between feminism and postmodernism', in Linda Nicholson, *Feminism/Postmodernism*, London: Routledge.

French, Marilyn (1978) *The Women's Room*, London: Sphere.

—— (1985) *Beyond Power, Women, Men and Morals*, London: Jonathan Cape.

—— (1992) *The War against Women*, London: Hamish Hamilton.

Freud, Sigmund (1935) *A General Introduction to Psychoanalysis*, London: Strachey.

Friedan, Betty (1963) *The Feminine Mystique*, New York: Dell, 1982.

Friend, John Kimball and Jessop, Norman (1976) *Local Government and Strategic Choice: An operational research approach to the processes of public planning*, London: Pergamon.

Gale, Andrew (1989a) 'Women in the British construction professions', *Gender and Science and Technology 5th International Conference*, Haifa, Jerusalem: GASAT.

—— (1989b) 'Attracting women to construction', *Chartered Builder*, September/ October, London: Chartered Institute of Building.

Gans, Herbert (1967) *The Levittowners*, London: Allen Lane.

Gardiner, A. (1923) *The Life of George Cadbury*, London: Cassell.

Gardner, Godfrey (1976) *Social Surveys for Social Planners*, Milton Keynes: Open University Press.

Gavron, Hannah (1966) *The Captive Wife*, London: Routledge.

Gaze, John (1988) *Figures in a Landscape: A history of the National Trust*, London: Barry & Jenkins, in association with the National Trust.

Geddes, Patrick (1915) *Cities in Evolution: An introduction to the town planning movement and to the study of civics*, London: Architectural Press, 1968.

—— and Thomson, J. Arthur (1889) *The Evolution of Sex*, London: Scott.

Gibbon, Gwilym (1942) *Reconstruction and Town and Country Planning*, London: The Architect and Building News.

Gibbs, Lesley (1987) 'Who designs the designers?', *WEB: Newsletter of Women in the Built Environment* 6: 4, July.

Gilligan, Carol (1982) *In a Different Voice: Psychological theory and women's development*, Cambridge, Mass.: Harvard University Press.

Gilman, Charlotte Perkins (1915) *Herland*, London: Women's Press, 1979.

—— (1921) 'Making towns fit to live in', *Century Magazine* 102: 361–6, July.

Girls' Own Paper (1904) 'The law relating to female suffrage', *Girls' Own Paper* XXVI (1304): 189, 24 December.

GLC (Greater London Council Women's Committee) (1984) *Working out the Future, women and jobs and the GLC*, London: GLC.

—— (1985) *Women on the Move: GLC survey on women and transport*, London: GLC.

—— (1986a) *Changing Places*, Report, London: GLC.

—— (1986b) 'Black and Ethnic Minority Women', *GLC Women's Committee* Issue 27, March, London: GLC.

—— (1986c) *Race and Planning Guidelines*, London: GLC.

Goffman, Erving (1969) *Presentation of Self in Everyday Life*, Harmondsworth: Penguin.

Goldsmith, Michael (1980) *Politics, Planning and the City*, London: Hutchinson.

Goldthorpe, John (1969) *Affluent Worker: Industrial attitudes and behaviour*, Cambridge: Cambridge University Press.

Goodison, Lucy (1990) *Moving Heaven and Earth: Sexuality, spirituality and social change*, London: Women's Press.

Goodwin, Barbara (1978) *Social Science and Utopia*, Sussex: Harvester.

Goret, Jean (1974) *La Pensée de Fourier*, Paris: Presses Universitaires de France.

Goss, Anthony (1965) *The Architect and Town Planning*, London: RIBA.

Gould, Peter and White, Rodney (1986) *Mental Maps*, London: Allen & Unwin.

Grant, Malcolm (1982 with 1990 supplement) *Urban Planning Law*, London: Sweet & Maxwell.

Grant, Malcomn (ed.) (1993) *Encyclopaedia of Planning*, London: Sweet & Maxwell.

Greed, Clara (1984) 'Whatever happened to patriarchal planning?', *Planning*, 8 June.

—— (1987) 'Drains feminism', letter in *Town and Country Planning*, Special Edition on 'A place for women in planning?', 56 (10): 279.

—— (1988) 'Is more better?: with reference to the position of women chartered surveyors in Britain', *Women's Studies International Forum*, 11 (3): 187–97.

—— (1990) 'The professional and the personal: a study of women in surveying', in Liz Stanley, (ed.) *Feminist Praxis: Research, theory and epistemology*, London: Routledge.

—— (1991) *Surveying Sisters: Women in a traditional male profession*, London: Routledge.

—— (1992a) 'The reproduction of gender relations over space: a model applied to the case of chartered surveyors', *Antipode: A Radical Journal of Geography* 24 (1): 16–28.

—— (1992b) 'Women chartered surveyors in Britain', *Architecture et Comportement*, 8 (2): 197–212, Lausanne: Association de la revue 'Architecture et comportement'.

—— (1992c) 'Women in planning', *The Planner* 78 (13): 11–13, 3 July.

—— (1992d) 'The changing position of women in planning', paper given at Scottish Branch of the RTPI, Carlton Highland Hotel, Edinburgh, 9 March.

—— (1993a) *Introducing Town and Country Planning*, London: Longmans.

—— (1993b) 'Is more better? Mark II: with reference to the position of women chartered surveyors in Britain', *Women's Studies International Forum* 16 (3): 255–70.

Greed, John (1978) *Glastonbury Tales*, Bristol: St Trillo Publications.

Green, Arthur (1986) *Jewish Spirituality*, London: Routledge.

Greer, Germaine (1973) *The Female Eunuch*, London: Paladin.

—— (1979) *The Obstacle Race: The fortunes of women painters and their work*, New York: Farrar Straus Giroux.

Greenwalt, Emmett (1955) *The Point Loma Community in California, 1897–1942: A theosophical experiment*, Berkeley: University of California Press.

Greenwich, London Borough of (1991) *Private Day Nurseries and Playgroups: Planning guidelines*, Greenwich: Design and Environment Section, Department of Planning.

Grieco, Margaret, Pickup, Laurie and Whipp, Richard (1989) *Gender, Transport and Employment*, Aldershot: Avebury.

Griffin, Susan (1984) *Women and Nature, The roaring inside her*, London: Women's Press.

Griffith, Ron and Amooquaye, Eno (1989) 'The place of race on the town planning agenda', *Planning Practice and Research* 4 (1).

Halford, Susan (1987) 'Women's initiatives in local government: tokenism or power?, *Working Paper 58*, Brighton: University of Sussex, Department of Urban and Regional Studies.

Halkes, Catherina (1991) *New Creation: Christian feminism and the renewal of the earth*, London: SPCK.

Hall, Peter (1977a) *Containment of Urban England*, London: Allen & Unwin.

—— (1977b) *The World Cities*, London: Weidenfeld & Nicolson.

—— (1980) *Great Planning Disasters*, London: Weidenfeld & Nicolson.

—— (1989) *Urban and Regional Planning*, London: Unwin Hyman.

Hall, Stuart and Jacques, Martin (1989) *New Times: The changing face of politics in the 1990's*, London: Lawrence & Wishart.

Hammersley, Martyn and Atkinson, Paul (1983) *Ethnography: Principles in practice*, London: Tavistock.

Hamnett, Chris, McDowell, Linda and Sarre, Philip (eds) (1989) *Restructuring Britain: The changing social structure*, London: Sage, in association with the Open University.

Hampson, Norman (1968) *The Enlightenment*, Harmondsworth: Penguin.

Hansard (1986) *Parliamentary Debates*, House of Lords, Official Report, Monday 16 June 1986, 476 (109–110): 120, reply by Lord Elton, London: HMSO.

Haraway, Donna (1989) *Primate Visions: Gender, race and nature in the world of modern science*, New York and London: Routledge.

—— (1991) *Simians, Cyborgs, and Women: The reinvention of nature*, London: Free Association Books.

Hardy, Dennis (1984) 'Lessons from Plotlands', *Town and Country Planning*, 17 November.

—— (1991a) *From Garden Cities to New Towns: Campaigning for town and country planning, 1899–1946*, London: Spon.

—— (1991b) *From New Towns to Green Politics: Campaigning for town and country planning, 1946–1990*, London: Spon.

Hargrove, Erwin and Conkin, Paul (eds) (1984) *TVA: The first fifty years of grass-roots bureaucracy*, Champaign: University of Illinois Press.

Hartmann, Heidi (1981) 'The unhappy marriage of marxism and feminism', in Lydia Sargent, (ed.) (1981) *Women and Revolution*, London: Pluto.

Harvey, David (1975) *Social Justice and the City*, London: Arnold.

Hass-Klau, Carmen, Nold, Inge, Böcker, Geert, and Crampton, Graham (1992) *Civilised Streets: A guide to traffic calming*, Brighton: Environmental and Transport Planning.

Hawkes, Nigel (1991) *Structures: Man-made wonders of the world*, London: Reader's Digest Association.

Hayden, Dolores (1976) *Seven American Utopias: The architecture of communitarian socialism*, Cambridge, Mass.: MIT.

—— (1981) *The Grand Domestic Revolution: Feminist designs for homes, neighborhoods and cities*, Cambridge, Mass.: MIT.

—— (1984) *Redesigning the American Dream*, London: Norton.

Healey, Patsy (1992) 'Town Planning in the 21st Century', conference paper, South Bank University, London.

—— McNamara, Paul, Elson, Martin and Doak, Andrew (1988) *Land Use Planning and the Mediation of Urban Change: The British planning system in practice*, Cambridge: Cambridge University Press.

Heap, Desmond (1991) *Outline of Planning Law*, London: Sweet & Maxwell.

Heine, Susanna (1987) *Women and Early Christianity*, London: SCM Press.

—— (1988) *Christianity and the Goddesses: Systematic criticism of a feminist theology*, London: SCM Press.

Heinen, Jacqueline (1992) 'Polish democracy is a masculine democracy', *Women's Studies International Forum*, 15 (1): 129–38.

Hennicker-Heaton (1964) *Day Release*, London: Ministry of Education.

Heron, Liz (ed.) (1985) *Truth, Dare or Promise: Girls growing up in the fifties*, London: Virago.

—— (1993) (ed.) *Streets of Desire: Women's fictions of the twentieth century*, London: Virago.

Herrington, John (1984) *The Outer City*, London: Harper & Row.

Hertz, Leah (1986) *The Business Amazons*, London: Andre Deutsch.

Herzenberg, Caroline, Meschel, Susan, and Altena, James (1991) 'Women scientists and physicians of antiquity and the Middle Ages', *Journal of Chemical Education* 68: 101, February.

Hewlett, Sylvia (1988) *A Lesser Life: The myth of women's liberation*, Harmondsworth: Penguin.

Hey, Valerie (1986) *Patriarchy and Pub Culture*, London: Tavistock.

Hill, William Thomson (1956) *Octavia Hill: Pioneer of the National Trust and housing reformer*, London: Hutchinson.

Hillman, Judy (1971) *Planning for London*, Harmondsworth: Penguin.

Hinchcliffe, Tanis (1988) 'Women as property owners', Paper given at the *Women in*

Planning History: Theories and applications seminar of the Planning History Group, York: Institute of Advanced Architectural Studies, 12 April.

Hinnells, Mark (1991) 'Not a load of rubbish – rationalising recycling schemes', *Town and Country Planning*, June: 181–3 (183=diagram).

Hirschon, Renee (ed.) (1984) *Women and Property, Women as Property: Power, property and gender relations*, London: Croom Helm.

Hite, Shere (1988) *The Hite Report, Women and Love: A Cultural revolution in progress*, London: Viking.

Hobhouse Report (1947) *Report of the National Parks Committee (England and Wales)*, London: HMSO.

Hochman, Elaine (1989) *Architects of Fortune*, New York: Fromm.

Hoggett, Brenda and Pearl, David (1983) *The Family, Law and Society*, London: Butterworths.

Holcombe, Lee (1983) *Wives and Property*, Oxford: Martin Robertson.

Holford, William (1949) 'Civic design: an inquiry into the scope and nature of town planning', *The Chartered Surveyor* 28 (8): 403–24, February.

Holloway, Richard (1991) *Who needs Feminism?: Men respond to sexism in the church*, London: SPCK.

Home, Robert (1993) 'Barrack camps for unwanted people: a neglected planning tradition', *Planning History*, 15 (1): 14–21.

Hoskins, John (1990) *Making of the English Landscape*, Harmondsworth: Penguin.

Howard, Ebenezer (1898) *Garden Cities of Tomorrow*, London: Faber & Faber, 1960 (ed F. J. Osborn).

Howatt, Hilary (1987) 'Women in planning – a programme for positive action', *The Planner* 73 (8): 11–12, August.

Howe, Elizabeth (1980) 'Role choices of urban planners', *Journal of the American Planning Association*, 398–409, October.

—— (1990) 'Normative ethics in planning', *Journal of Planning Literature* 5 (2).

—— and Kaufman, Jerome (1981) 'The values of contemporary American planners', *Journal of the American Planning Association* , 266–78 July.

Hudson, Mike (1978) *The Bicycle Planning Book*, London: Open Books, with Friends of the Earth.

Hutchinson, Max (1990) 'Heroes', *Architects Journal*, 14 February: 5.

IJURR (International Journal of Urban and Regional Research) (1978) *Women and the City*, Special Issue, 2 (3), London: Basil Blackwell.

Ingham, Mary (1981) *Now We are Thirty: Women of the breakthrough generation*, London: Methuen.

Jacobs, Jane (1964) *The Death and Life of Great American Cities: The failure of town planning*, Harmondsworth: Penguin.

—— (1970) *The Economy of Cities*, Harmondsworth: Penguin.

Jagger, Alison (1983) *Feminist Politics and Human Nature*, Brighton: Harvester.

Jeffreys, Sheila (1990) *Anticlimax: A feminist perspective on the sexual revolution*, London: Women's Press.

JFCCI (Joint Forecasting Committee for the Construction Industries) (1991) *Construction Forecasts*, London: National Economic Development Office.

Jones, Greta (1986) *Social Hygiene in Twentieth Century Britain*, London: Croom Helm.

Joseph, Martin (1978) 'Professional values, a case study of professional students in a Polytechnic', *Research in Education*, 19: 49–65, May.

Kanter, Rosabeth Moss (1972) *Commitment and Community: Communities and utopias in sociological literature*, Cambridge, Mass.: Harvard University Press.

—— (1977) *Men and Women of the Corporation*, New York: Basic Books.

Keeble, Lewis (1969) *Principles and Practice of Town and Country Planning*, London: Estates Gazette.

—— (1983) *Town Planning Made Plain*, London: Longman.

Keller, Suzanne (1981) *Building for Women*, Cambridge, Mass.: Lexington Books.

Kent, Susan Kingsley (1987) *Sex and Suffrage in Britain: 1860–1914*, London: Routledge.

Kessler, Carol Farley (1984) *Daring to Dream: Utopian stories by United States women, 1836–1919*, Boston: Pandora.

Kimball-Hubbard, Theodora and Hubbard, Henry (1929) *Our Cities Today and Tomorrow: A survey of planning and zoning programs in the United States of America*, Cambridge, Mass.: Harvard University Press.

King, Anthony (1984) *The Bungalow: The production of a global culture*, London: Routledge.

King, Graham (1991) 'The darling buds of May', *The Planner* 77 (16): 6, 10 May.

—— (1992) 'The City as Matrix' (book review of Sizemore, 1989) *Town and Country Planning* 61 (4): 123–4, April.

King, Michael, Israel, Mark and Goulbourne, Selina (1990) *Ethnic Minorities and Recruitment to the Solicitors' Profession*, London: Law Society.

Kirk, Gwyneth (1980) *Urban Planning in a Capitalist Society*, London: Croom Helm.

Kirkup, Gill and Keller, Laurie Smith (1992) *Inventing Women: Science, gender and technology*, Cambridge: Polity Press, in association with the Open University.

Kitchen, Ted (1970) 'Planning education in Britain: The state of the game', *The Planner* 56: 20–2 and 611–12.

Kitzinger, Celia (1991) 'Sex, beauty and beasts', *New Internationalist*, January, 215: 18–19 (issue on animals rights).

Knappert, Jan (1990) *The Aquarian Guide to African Mythology*, London: Aquarian Press.

Knight, Stephen (1985) *The Brotherhood*, London: Panther.

Knox, Paul (ed.) (1988) *The Design Professions and the Built Environment*, London: Croom Helm.

Kolmerton, Carol (1990) *Women in Utopia: The ideology of gender in the American Owenite communities*, Bloomington: Indiana University Press.

Korda, Michael (1974) *Male Chauvinism: How it works*, London: Barrie & Jenkins.

Krishnarayan, Vijay (1990) *Ethnic Minorities and the Planning System*, London: Royal Town Planning Institute.

Kumar, Krishan (1991) *Utopianism*, Milton Keynes: Open University Press.

Lake, Brian (1941) 'A plan for Britain', special issue of *Picture Post*, 10 (1), (Special Issue No 7, 1974) London: Peter Way Ltd.

Lamplugh, Diana (1988) *Beating Aggression: A practical guide for working women*, London: Weidenfeld & Nicolson.

Landau, Cécile (1991) *Growing up in the Sixties*, London: Optima.

Lane, Michael (1975) *Design for Degrees: A history of CNAA from its inception in 1964*, London: Macmillan.

Langland, Elizabeth and Gove, Walter (eds) (1981) *A Feminist Perspective in the Academy: the difference it makes*, Chicago: University of Chicago Press.

Lateef, Shahida (1990) *Muslim Women in India: Politial and private realities*, London: Zed Books.

Lawless, Paul (1981) *Britain's Inner Cities*, London: Macmillan.

Law Society (1988) *Equal in the Law: Report of the Working Party on Women's Careers*, London: The Law Society.

LBDRT (London Boroughs Disability Resource Team) (1991) 'Access for all: Special feature', *Community Network*, 8, (3), Autumn, London: Town and Country Planning Association.

Leach, Penelope (1979) *Who Cares? A new deal for mothers and their small children*, Harmondsworth: Penguin.

Leavitt, Jacqueline (1981) 'The History, status, and concerns of women planners', in Catherine Stimpson, Elsa Dixler, Martha Nelson, and Kathryn Yatrakis, (eds) (1981) *Women and the American City*, Chicago: University of Chicago Press.

Lederman, Alfred and Trachsel, Alfred (1959) *Playground and Recreational Spaces*, New York: Architectural Press.

Lees, Terence (1979) *Psychology and the Environment*, London: Methuen.

Leevers, Kate (1986) *Women at Work In Housing*, London: HERA.

Legrand, Jacques (1988) *Chronicle of the Twentieth Century*, London: Chronicle.

LeGuin, Ursula (1969) *The Left Hand of Darkness*, London: Orbit.

Leoff, Constance (1987) *Bluff Your Way in Feminism*, London: Ravette.

Levin, Ira (1974) *Stepford Wives*, London: Pan.

Levine, Philippa (1987) *Victorian Feminism: 1850–1900*, London: Hutchinson.

—— (1990) *Feminist Lives in Victorian England*, London: Blackwell.

Levison, Debra and Atkins, Julia (1987) *The Key to Equality: The women in housing survey*, Women in Housing Working Party, London: Institute of Housing.

Lewis, Jane (1984) *Women in England: 1870–1950*, Brighton, Sussex: Harvester.

—— (1992a) *Women in England: since 1945*, Oxford: Blackwell.

—— (1992b) 'Comment', *Signs* 18 (1): 203.

Lewis, Richard and Talbot-Ponsonby, Andrew (1986) *The People, the Land and the Church*, Hereford: Hereford Diocesan Board of Finance.

LFHG (London Feminist History Group) (1983) *The Sexual Dynamics of History: Men's power, women's resistance*, London: Pluto.

LGMB (Local Government Management Board) (1992) *Summary: Survey of planning staffs in local authority planning departments as at 31st October, 1991*, London: LGMB with RTPI.

Lichfield, Nathaniel (1975) *Evaluation in the Planning Process*, London: Pergamon.

Lindsay, Jean (1993) *Elizabeth M. Mitchell: The happy town planner*, Edinburgh: The Pentland Press.

Lipovskaya, Olga (forthcoming) 'The post-communist woman: sisters and cousins: how close is sisterhood?', in Rosalind Marsh (ed.) *Women in Russia and the Former USSR*, Cambridge: Cambridge University press, forthcoming.

Little, Jo (1990) 'Rural idyll? Planning for women in rural areas', *WEB, Women and the Built Environment*, 15/16 : 24–25, Autumn.

—— Peake, Linda and Richardson, Pat (1988) *Women and Cities, Gender and the urban environment*, London: Macmillan.

—— (1991) 'Political economy working paper: is gender important in planning?' (unpublished).

—— (1994) *Gender, Planning and the Policy Process*, London: Pergamon.

Lockwood, W. B. (1965) *An Informal History of the German Language: With chapters on Dutch, Afrikaan, Frisian and Yiddish*, Cambridge: Heffer.

Lorber, Judith (1984) *Women Physicians*, London: Tavistock.

Lorenz, Clare (1990) *Women in Architecture: A contemporary perspective*, London: Trefoil.

LPAS (London Planning Aid Service) (1986a) *Planning for Women: An evaluation of local plan consultation by three London boroughs*, Research Report 2, London: TCPA.

—— (1986b) *Planning Advice for Women's Groups*, Community Manual 6, London: TCPA.

Luomala, Nancy (1982) 'Matrilineal reinterpretation of some Egyptian sacred cows', in Norma Broude and Mary D. Garrard, *Feminism and Art History*, New York: Harper & Rowe.

Lutyens, Mary (1980) *Edwin Lutyens: by his daughter*, London: John Murray.

LWPG (London Women and Planning Group) (1991) *Shaping our Borough: Women and unitary development plans*, London: Planning Aid for London.

Lynch, Kevin (1960) *The Image of the City*, Cambridge, Mass. and London: MIT.

Mabin, Alan (1991) 'Origins of segregatory urban planning in South Africa c. 1900–1940', in *Planning History* 13 (3): 8–16, University of Witwatersrand, Johannesburg, South Africa.

McAuslan, Patrick (1980) *Ideologies of Planning Law*, London: Pergamon.

McBride, Theresa (1976) *The Domestic Revolution: The modernization of household service in England and France*, New York: Holmes & Meier.

McCaffrey, Anne (1990) *The Rowan*, London: Corgi.

McCall, Cicely (1958) *Our Villages*, London: Women's Institute.

McCallum, Ian (1946) *Physical Planning: The ground work of a new technique*, London: Architectural Press.

McDowell, Linda (1983) 'Towards an understanding of the gender division of urban space', *Environment and Planning D: Society and Space*, 1: 59–72.

—— (1986) 'Beyond patriarchy: a class-based explanation of women's subordination', *Antipode: A Radical Journal of Geography* 18 (3): 311–21.

—— (1991) 'Restructuring production and reproduction: some theoretical and empirical issues relating to gender, or women in Britain', in M. Gottdiener and C. Pickvance (eds) *Urban Life in Transition*, 39, Urban Affairs Annual Reviews (book series), London: Sage.

—— (1992) 'Multiple voices: speaking from outside and inside "the Project"', *Antipode: A Radical Journal of Geography*, 24 (1), January.

—— and Pringle, Rosemary (eds) (1992) *Defining Women: Social institutions and gender divisions*, Milton Keynes: Open University, in association with Blackwells, Oxford.

MccGwire, Scarlett (1992) *Best Companies for Women: Britain's top employers*, London: Harper-Collins, Pandora.

MacKeith, Margaret (1986) *Shopping Arcades: A gazetteer of extant British arcades, 1817–1939*, London: Mansell.

MacKenzie, Norman and Jeanne (1986) *The Diary of Beatrice Webb*, London: Virago.

MacKenzie, Suzanne (1989) *Visible Histories: Women and environments in a post-war British city*, Montreal: McGill-Queen's University Press.

McLaren, Angus (1978) *Birth Control in Nineteenth Century England*, New York: Holmes & Meier.

Macleod, Iain (1991) *The Competence of an Ingenieur*, Department of Civil Engineering, University of Strathclyde.

McLoughlin, J. (1969) *Urban and Regional Planning: A system's view*, London: Faber.

McMahon, Michael (1985) 'The Law of the land: property rights and town planning in modern Britain', in M. Ball (ed.), *Land Rent, Housing and Urban Planning*, London: Croom Helm.

McQuiston, Liz (1989) *Women in Design*, London: Trefoil.

Mahoney, Pat (1985) *Schools for the Boys?* London: Hutchinson.

Malinowski, Bronislav (1977) *Freedom and Civilisation*, London: Greenwood Press.

Malpass, Peter (1975) 'Professionalism and the role of architects in local authority housing', *Royal Institute of British Architects Journal* 82.

Malthus, Thomas (1798) *Essay on the Principles of Population*, London: Dent, 1973.

Mamonova, Tatyana (1989) *Russian Women's Studies: Essays on sexism in Soviet culture*, London: Pergamon.

Manchester (1987) *Planning a Safer Environment for Women*, Manchester: Planning for Women Group, Manchester City Planning Department.

Manuel, Frank and Fritzie (1979) *Utopian Thought in the Western World*, Oxford: Basil Blackwell.

Mann, Jean (1962) *Women in Parliament*, London: Odhams.

Mansbridge, Jane (1986) *Why we Lost the ERA*, London: University of Chicago Press.

Marcus, Susanna (1971) 'Planners – who are you?' *Journal of the Royal Town Planning Institute* 57 (2).

Markale, Jean (1986) *Women of the Celts*, Rochester, Vermont: Inner Traditions International (trans. from French).

Marks, Peter (1988) *Solicitors' Career Structure Survey*, London: Polytechnic of Central London and Law Society.

Markusen, Anne (1981) 'City spatial structure, women's household work and national urban policy' in Catherine Stimpson, Elsa Dixler, Martha Nelson, and Kathryn Yatrakis, (eds) *Women and the American City*, Chicago: University of Chicago Press.

Marriot, Oliver (1967) *The Property Boom*, London: Pan, 1989.

Marshall, Judi (1984) *Women Managers: Travellers in a male world*, London: Wiley.

Marwick, Arthur (1986) *Class in the Twentieth Century*, Brighton: Harvester.

Marx, Karl (1857) (trans. A. Miller) *Grundrisse*, Harmondsworth: Penguin, 1981.

Massey, Doreen (1984) *Spatial Divisions of Labour: Social structures and the geography of production*, London: Macmillan.

—— and Allen, John (eds) (1984) *Geography Matters*, Cambridge: Cambridge University Press, in association with the Open University.

Massingham, Betty (1984) *Miss Jekyll: Portrait of a great gardener*, Newton Abbott: David & Charles.

Matless, David (1992) 'Regional surveys and local knowledges: the geographical imagination of Britain, 1918–39' in *Transactions*, 17 (4): 464–80, London: Institute of British Geographers.

MATRIX (1984) *Making Space: Women and the man made environment*, London: Pluto.

Maxwell, Margaret (1990) *Narodoniki Women: Russian women who sacrificed themselves for the dream of freedom*, London: Pergamon.

Mazumdar, Pauline (1991) *Eugenics, Human Genetics and Human Failings: The Eugenics Society, its sources and its critics in Britain*, London: Routledge.

Mead, Margaret (1949) *Male and Female*, London: Gollancz.

—— (1966) *Coming of Age in Samoa*, Harmondsworth: Pelican.

—— (1968) 'Houses are Homes', in Stephen Verney, *People and Cities*, Coventry: Conference Report, Coventry Cathedral.

Merchant, Vicki (1993) *Draft Harassment Policy*, Preston: University of Central Lancashire, Equal Opportunities Committee.

Merton, Robert (1952) 'Bureacratic structure and personality', in Richard Merton, (ed.) *Reader in Bureaucracy*, New York: Free Press.

Metcalfe, Andy and Humphries, Martin (eds) (1985) *The Sexuality of Men*, London: Pluto.

Midwinter, Eric (1972) *Priority Education*, Harmondsworth: Penguin.

Mies, Maria (1987) 'Why do we need all this? A call against genetic engineering and reproductive technology', in Patricia Spallone and Deborah Steinberg (1987) *Made to Order: The myth of reproductive and genetic progress*, Oxford: Pergamon.

Miles, Rosalind (1988) *The Women's History of the World*, London: Paladin.

Miller, Daniel and Swanson, Guy (1958) *The Changing American Parent: A study in the Detroit area*, New York: Wiley.

Miller, Mervyn (1991) 'Homesgarth: Howard's model for co-operative living', in *Town and Country Planning* 60 (4): 119–120 April.

Millerson, Geoffrey (1964) *The Qualifying Associations*, London: Routledge & Kegan Paul.

Millett, Kate (1970) *Sexual Politics*, London: Virago, 1985.

Ministry of Education (1966), *A Plan for Polytechnics and other Colleges*, White Paper, London: HMSO.

Mishan, E. J. (1973) *Cost Benefit Analysis*, London: George Allen & Unwin.

Mitchell, Elizabeth B. (1967) *The Plan that Pleased*, London: Town and Country Planning Association.

Mitchell, Juliet and Oakley, Anne (eds) (1986) *What is Feminism?*, Oxford: Blackwell.

Mohney, David and Easterbury, Keller (1991) *Seaside: Making of a town in America*, London: Phaidon.

Momsen, Janet and Townsend, Janet (1987) *Gender of Geography in the Third World*, London: Hutchinson.

Montgomery, John and Thornley, Andy (1990) *Radical Planning Initiatives: New directions for urban planning*, Aldershot: Gower.

Moody, Nickianne (1991) 'Maeve and Guinevere: women's fantasy writing in the science fiction marketplace', in Lucie Armitt (1991) *Where No Man has Gone Before: women and science fiction*, London: Routledge.

Moore, Robert (1977) 'Becoming a sociologist in Sparkbrook', in Colin Bell and Howard Newby (eds) *Doing Sociological Research*, London: George Unwin.

Moore, Victor (1987 with 1991 supplement) *A Practical Approach to Planning Law*, London: Blackstone.

More, Thomas (1516) *Utopia*, London: Cambridge University Press, 1989.

Morgan, Elaine (1974) *The Descent of Woman*, London: Corgi.

—— (1978) *Falling Apart: the rise and decline of urban civilisation*, London: Abacus.

Morgan, Peter and Nott, Susan (1988) *Development Control: Policy into practice*, London: Butterworths.

Morphet, Janice (1983) 'Planning and the majority – Women', paper given at The Town and Country Planning School, St Andrews, Scotland, London: The Royal Town Planning Institute.

Morris, A. E. J. (1972) *History of Urban Form: Prehistory to the Renaissance*, London: George Godwin.

Morris, Anne and Nott, Susan (1991) *Working Women and the Law: Equality and discrimination in theory and practice*, London: Routledge.

Morris, Eleanor (1986) 'An overview of planning for women, 1945–75', in Conference Report, *Women and Planning in Scotland*, Linlithgow, West Lothian: RTPI Scotland.

Morris, Meaghan (1988) *The Pirate's Fiancée: Feminism reading postmodernism*, London: Verso.

Morris, Terence (1958) *The Criminal Area*, London: Routledge & Kegan Paul.

Moser, Caroline (1993) *Gender, Planning and Development*, London: Routledge.

—— and Peake, Linda (1987) *Women, Human Settlements and Housing*, London: Tavistock.

Mulford, Wendy (1986) 'In this process, I too am subject', in Denise Farran, Sue Scott, and Liz Stanley (eds) *Writing Feminist Bioiography*, Manchester: University of Manchester.

Mumford, Lewis (1930s) *The City*, American Institute of Planners film, by RKO, commentary written by Mumford, shown 8 August 1987, on BBC2.

—— (1938) *Culture of Cities*, New York: Free Press.

—— (1945) *City Development: Studies in disintegration and renewal*, Westport, Conn.: Greenwood Press, 1973.

—— (1965) *The City in History*, Harmondsworth: Penguin.

Musgrove, Anne (1992) 'Think Tank: Planning for the Future' *ITA Magazine*, Capricorn Publishing, 78 George Street, NSW: 20–4, January.

Myerscough, Cyril (1975) *Feet First: A pedestrian survival handbook*, London: Wolfe, and the Pedestrians' Association for Road Safety.

Nadin, Vincent and Jones, Sally (1990) 'A Profile of the Profession', *The Planner* 76 (3): 13–24, 26 January.

Neale, John (1961) *Queen Elizabeth I*, Harmondsworth: Penguin.

Nethercot, Arthur (1963) *The Last Four Lives of Annie Besant*, Chicago: University of Chicago Press.

Nettlefold, J. S. (1910) *Practical Housing*, London: Garden City Press.

New Internationalist (1992) 'We've only just begun: feminism in the 1990s', *New Internationalist* 227, January, whole issue.

Newby, Howard (1982) *Green and Pleasant Land*, Harmondsworth: Penguin.

Newman, Oscar (1973) *Defensible Space: People and design in the violent city*, London: Architectural Press.

Newsom, Carol and Ringe, Sharon (1992) *The Women's Bible Commentary*, London: SPCK.

Nicholson, Linda (ed.) (1990) *Feminism/Postmodernism*, London: Routledge.

Nietsche, Friederich (1973) *Will to Power: In science, nature, society and art*, London: Vintage Books.

Nin, Anaïs (1959) *Cities of the Interior*, London: Peter Owen, 1978.

Norris, June (1961) *THe Human Aspects of Redevelopment*, Birmingham: Midlands New Towns Society.

Norton-Taylor, Richard (1982) *Whose Land is it Anyway?*, Wellingborough: Turnstone.

Nott, Susan (1989) 'Women in the law', *New Law Journal* 139, 6410: 749–52, continued 139, 6411: 785–6, 2–7 June.

Nuttgens, Patrick (1972) *The Landscape of Ideas*, London: Faber & Faber.

Oakley, Anne (1980) *Women Confined: Towards a sociology of childbirth*, Oxford: Martin Robertson.

Oakley, Anne (1992) *Social Support and Motherhood*, London: Routledge.

Okin, Susan, Moller (1979) *Women in Western Political Thought*, London: Virago.

Oliver, P., Davis, I. and Bentley, I. (1981) *Dunroamin: The suburban semi and its enemies*, London: Barrie & Jenkin.

OPCS (Office of Population Censuses and Surveys) (1983 update) 'Usually resident population: age by marital status by sex', Table 6, p. 15, *Census 1981: National Report – Great Britain, Part I*, London: HMSO.

—— (1990) *Standard Occupational Classification*, Registrar General, London: HMSO.

—— (1991) *Social Trends*, London: HMSO.

Osen, Lynn (1974) *Women in Mathematics*, Cambridge, Mass.: MIT.

Owen, Alex (1989) *The Darkened Room: Women, power and spiritualism in late Victorian England*, London: Virago.

Owen, Robert (1849) *The Book of the New Moral World: Containing the rational system of society*, London: J. Watson.

Pahl, Ray (1965) *Urbs in Rure*, London: Weidenfeld & Nicolson.

—— (1977) 'Managers, technical experts and the state', in M. Harloe (ed.) *Captive Cities*, London: Wiley.

—— (1984) *Divisions of Labour*, Oxford: Blackwell.

—— and Pahl, Jan (1971) *Managers and their Wives*, Harmondsworth: Penguin.

Pain, Gillian (1967) *Planning and the Shopkeeper: With particular reference to the problem of commercial deliveries to and from shops*, London: Barrie & Rockcliff.

Papafio, Kwarley (1991) 'Black women in housing', *HERA Quarterly* 125: 1, March.

Pardo, Vittorio (1965) *Le Corbusier*, London: Thames & Hudson.

Parker Morris Report (1961) *Homes for Today and Tomorrow*, London: Central Housing Advisory Committee.

Parkin, Frank (1979) *Marxism and Class Theory: A bourgeois critique*, London: Tavistock.

Parrington, Vernon (1930) *Main Currents in American Thought*, New York: Free Press.

Pateman, Carole (1992) 'The patriarchal welfare state' in Linda McDowell and Rosemary Pringle (eds) *Defining Women: Social institutions and gender divisions*, Oxford: Polity Press, in association with the Open University.

Patterson, Richard (1992) 'Discourse and assessment' (unpublished) CISC Workshop, London.

Pearsall, Phyllis (1990) *From Bedsitter to Household Name: The personal story of A-Z maps*, London: Geographers' A-Z Company.

Pearson, Lynn (1988) *The Architectural and Social History of Cooperative Living*, London: Macmillan.

Peckham, M. (1970) *The Triumph of Romanticism; Collected essays*, Charlotte: University of South Carolina Press.

Percy Committee (1970) *Report of the Committee on Higher Technical Education*, London: HMSO.

Pevsner, Nikolaus (1970) *Pioneers of Modern Design*, Harmondsworth: Penguin.

Phillips, Patricia (1990) *The Scientific Lady: A social history of woman's scientific interests 1520–1918*, London: Weidenfeld & Nicolson.

Pickup, Laurie (1984) 'Women's gender role and its influence on their travel behaviour', in *Built Environment*, Special Issue on 'Women and the built environment', 10, (1): 61–8.

Pickvance, Christopher, (ed.) (1977) *Urban Sociology*, London: Tavistock.

Piercy, Marge (1979) *Woman on the Edge of Time*, New York: Fawcett Press.

Pinch, Stephen (1985) *Cities and Services: The geography of collective consumption*, London: Routledge & Kegan Paul.

Pizan, Christine de (1405)(trans. E. J. Richards) *The Book of the City of Ladies*, New York: Persea Books, 1982.

Plato (384 BC) *Republic*, London: J. M. Dent, 1926.

Popper, Karl (1958) *The Poverty of Historicism*, London: Routledge.

Poulton, K. and Hunt, L. (1986) 'Planning for women: An evaluation of consultation in three London boroughs', *London Planning Aid Service: Research Paper 2*, London: Town and Country Planning Association.

Power, Anne (1987) *Property Before People: The management of twentieth-century council housing*, London: Allen & Unwin.

Priest, Gordon and Cobb, Pamela (1980) *The Fight for Bristol: Planning and the growth of public protest*, Bristol: Bristol Civic Society and Redcliffe Press.

Priestley, J. B. (1934) *English Journey*, London: Heinemann, 1984.

Prince of Wales (1989) *A Vision of Britain*, London: Doubleday.

Punter, John (1990) *Design Control in Bristol: 1940–1990*, Bristol: London Press.

Purvis, June (1991) *A History of Women's Education in England*, Milton Keynes: Open University.

Rakodi, Carol (1991) 'Planning for women in developing countries', unpublished talk to the Woman and Planning, London Boroughs Meeting, 17 April.

—— and Mutizwa-Mangiza, N. D. (1989) *Housing Policy, Production and Consumption: A case study of Harare*, RUP: University of Zimbawbe.

Ramazanoglu, Caroline (1992) 'Feminism and liberation', in L. McDowell and Rosemary Pringle (eds) *Defining Women: Social institutions and gender divisions*, Cambridge: Polity, in association with the Open University.

Rao, N. (1990) *Black Women in Public Housing*, London: London Race and Housing Unit, Brixton.

Rapoport, Rhona and Rapoport, Robert (1971) *Dual Career Families*, Harmondsworth: Penguin.

Ravetz, Alison (1974) *Model Estate: Planned housing at Quarry Hill, Leeds*, London: Croom Helm, in association with the Joseph Rowntree Memorial Trust.

—— (1980) *Remaking Cities: Contradictions of the recent urban environment*, London: Croom Helm.

—— (1986) *The Government of Space: Town planning and modern society*, London: Faber & Faber.

—— (1989) 'A view from the interior' in J. Attfield and P. Kirkham (eds) *A View from the Interior: Feminism, design and women*, London: Women's Press.

—— (1993) 'Talking houses: but what of house design', *Town and Country Planning*, 62 (4): 85, April.

Ratcliffe, John (1981) *Introduction to Town and Country Planning*, London: Hutchinson.

Reade, Eric (1987) *British Town and Country Planning*, Milton Keynes: Open University Press.

Reed, Rosslyn (1990) 'Compositor: a job fit for a woman?', paper presented at Ninth Annual International UMIST–Aston Labour Process Conference on the Organisation and Control of the Labour Process, Manchester: University of Manchester.

Rees, Gareth and Lambert, John (1985) *Cities in Crisis*, London: Arnold.

Rees, Nigel (1979) *Graffiti Lives, O.K.*, London: Unwin.

Reith Report (1946) *New Towns Committee: Interim Report; Final Report*, London: HMSO.

Rex, John and Moore, Robert (1967) *Race, Community and Conflict*, London: Institute of Race Relations.

Richardson, B. (1876) *Hygenia: A City of health*, London.

RICS (Royal Institution of Chartered Surveyors) (1990) 'Report of the Working Party on Equal Opportunities', London: RICS (internal document), and (published from this) *Model Policy Statement for Equal Opportunities for Members' Firms*, London: RICS.

Riley, Mary and Bailey, Christine (1983) 'Learning to get by in a man's world', *Planning* 543, 4 November.

Roberts, Helen (1981) *Doing Feminist Research*, London: Routledge.

Roberts, Margaret (1974) *Town Planning Techniques*, London: Hutchinson.

Roberts, Marion (1991) *Living in a Man-made World: Gender assumption in modern housing design*, London: Routledge.

Roberts, Patricia (1988a) 'The image of women in post war plans for London', Conference Paper presented at York on *Women and Planning History: Theories and applications*, London: South Bank Polytechnic.

—— (1988b) 'Women and planning history: theories and applications', Paper given at *Women and Planning History: Theories and applications*, Seminar of the Planning History Group, York: Institute of Advanced Architectural Studies.

Robbins Report (1963) *Higher Education*, Ministry of Education, London: HMSO.

Robinson, Eric (1968) *The New Polytechnics*, Harmondsworth: Penguin.

Rodriguez-Bachiller, Augustin (1988) *Town Planning Education*, Aldershot: Avebury/Gower.

Rogers, Barbara (1980) *The Domestication of Women: Discrimination in developing societies*, London: Tavistock.

—— (1983) *52 per cent: Getting Women's Power into Politics*, London: Women's Press.

—— (1988) *Men Only: An investigation into men's organisations*, London: Pandora.

Rose, Gillian (1993) *Feminism and Geography: The limits of geographical knowledge*, Cambridge: Polity.

Rosenau, Helen (1983) *The Ideal City: Its architectural evolution in Europe*, London: Methuen.

Rosser, Sue (1992) 'Are there feminist methodologies appropriate for the natural sciences and do they make a difference?', *Women's Studies International Forum*, 15 5/6.

Rossi, William (1977) *The Sex Life of the Foot and Shoe: An occasionally indecent exploration of the secret history of feet and footwear*, London: Routledge & Kegan Paul.

RTPI (Royal Town Planning Institute) (1971) *Town Planners and their Future: A discussion paper*, London: RTPI.

—— (1983) *Planning for a Multi-Racial Britain*, London: CRE, Commission for Racial Equality.

—— (1984) *Sample Survey of Members 1984: Interim results*, London: RTPI.

—— (1986) *Planning History*, Planning course 1, Block 1, Units 1–5, Distance Learning Course, London: RTPI, with Bristol and Leeds Polytechnics.

—— (1987) *Report and Recommendations of the Working Party on Women and Planning*, London: RTPI.

—— (1988) *Managing Equality: the role of senior planners*, Conference, London: RTPI, 28 October.

—— (1989a) *Planning for Town and Country, Context and Achievement, 1914–1989: 75th anniversary brochure of the Royal Town Planning Institute*, London: RTPI.

—— (1989b) *Planning for Choice and Opportunity*, Papers prepared by the Women and Planning Working Party, London: RTPI.

—— (1990) *Careers in Town Planning*, London: RTPI.

—— (1991) *Traffic Growth and Planning Policy*, London: RTPI.

—— (1992) 'Guidance Note B', annex to *The Education of Planners: Policy statement and general guidance for academic institutions offering initial professional education in planning*, London: RTPI.

—— (1993) *Ethnic Minorities and the Planning System*, London: RTPI.

—— South East Branch (1990) Women's Network Meeting, Crawley, Convenor Colette Blackett.

—— South West Branch (1991) Traffic and Transport; Alternative Strategies, meeting Guildhall, Bath, 4 October.

—— South West Branch (1992) *Effective Communication*, Taunton, Conference Report, 18 June.

—— Yorkshire Branch (1986) *Report of a Questionnaire Survey of Women in Planning*, London: RTPI.

—— Yorkshire Branch (1991a) *Flexible Working Seminar*, (Sheffield and Doncaster Women and Planning Working Group, Conference, 1 February).

—— Yorkshire Branch (1991b) *Equity in the City: A positive approach to planning*, (Sheffield and Doncaster, Women and Planning Working Group, Conference February).

—— (1993) *Ethnic Minorities and the Planning System*, London: RTPI.

Rubenstein, W. D. (1981) *Men of Property: The very wealthy in Britain since the Industrial Revolution*, London: Croom Helm.

Russell, Elizabeth (1983) 'Utopian dreams and dystopian nightmares: a general survey of utopias written by women in English 1792–1937', unpublished Ph.D. thesis, University of Barcelona, Department of English.

Russell, John E. (1954) *World Population and World Food Supplies*, London: George Allen & Unwin.

Rydin, Yvonne (1993) *The British Planning System: An introduction*, London: Macmillan.

Saffioti, Heleieth (1978) *Women in Class Society*, New York: Monthly Review Press.

Saks, Mike (1983) 'Removing the blinkers: A critique of recent contributions to the sociology of the professions', *Sociological Review*, 2–21, February.

Sandercock, Leonie and Forsyth, Ann (1992) 'A gender agenda: new directions for planning theory', *Journal of the American Planning Association*, 58, (1): 49–58, Winter.

Saunders, Peter (1979) *Urban Politics: A sociological interpretation*, Harmondsworth: Penguin.

—— (1985) 'Space, the city and urban sociology', in Derek Gregory and John Urry (1985) *Social Relations and Spatial Structures*, London: Macmillan.

—— and Williams, Peter (1988) 'The constitution of the home: towards a research agenda', *Housing Studies* 3, 2.

Savage, Wendy (1986) *A Savage Enquiry: Who controls childbirth?*, London: Virago.

SBP (South Bank Polytechnic) (1987) *Women and their Built Environment*, Conference in the Faculty of the Built Environment, South Bank Polytechnic, London: SBP.

Scarman, Lord (1982) *The Scarman Report: The Brixton disorders, 10–12 April 1981*, Harmondsworth: Penguin.

Schreiner, Olive (1911) *Woman and Labour*, London: Virago, 1978.

Schuster Committee (1950) *Report on the Qualifications of Planners*, Cmd. 8059, London: HMSO.

Scott, N. K. (1989) *Shopping Center Design*, New York: Van Nostrand Reinhold.

Scott Report (1942) *Report of the Committee on Land Utilisation in Rural Areas*, London: HMSO.

Sebestyen, Amanda (1988) *'68, '78, '88: From Women's Liberation to Feminism*, Bridport, Dorset: Prism Press.

Segal, Lynne (1987) *Is the Future Female?: Troubled thoughts on contemporary feminism*, London: Virago.

Sertima, Ivan Van (1985) *Egypt Revisited*, London: Transaction.

Service, Alastair (1977) *Edwardian Architecture*, London: Thames & Hudson.

Sharp, Thomas (1931) *Town and Countryside: Some aspects of urban and rural development*, Oxford: Oxford University Press.

Sharpe, Sue (1976) *Just like a Girl: How girls learn to be women*, Harmondsworth: Penguin.

Shaw, Sue (1991) 'Test tube coup: biotech's global takeover', *New Internationalist* 217: 4–5, March.

Sheffield (1986) *Central Area Local Plan*, Sheffield: Sheffield City Planning Department.

Shelley, Mary Wollstonecraft (1818) *Frankenstein, or, The Modern Prometheus*, Oxford: Oxford University Press, 1980.

Shoard, Marion (1980) *The Theft of the Countryside*, London: Temple Smith.

—— (1987) *This Land is our Land: The struggle for Britain's countryside*, London: Paladin.

Showalter, Elaine (1982) *A Literature of their Own: From Charlotte Brontë to Doris Lessing*, London: Virago.

Siltanan, Janet and Stanworth, Michelle (1984) *Women and the Public Sphere*, London: Hutchinson.

Silverstone, Rosalie and Ward, Audrey (ed.) (1980) *Careers of Professional Women*, London: Croom Helm.

Simmie, James (1974) *Citizens in Conflict, The sociology of town planning*, London: Hutchinson.

—— (1981) *Power, Property and Corporatism*, London: Macmillan.

Sitte, Camillo (1889) *Der Städtebau: Nach Seinen Künstlerishen Grundsätzen* (*City Planning According to Artistic principles*) Vienna.

Sizemore, Christine Wick (1989) *The Female Vision of the City*, Memphis: University of Tennessee Press.

Sjoberg, Gidean (1965) *Pre-Industrial City: Past and present*, New York: Free Press.

Skeffington, A.(1969) *People and Planning*, London: HMSO.

Smart, Carol (1984) *The Ties that Bind: Law, marriage and the reproduction of patriarchal relations*, London: Routledge & Kegan Paul.

Smith, Neil and Williams, Peter (1986) *Gentrification of the City*, London: Allen & Unwin.

Smith, Susan (1989) *The Politics of Race and Residence*, Oxford: Polity.

SNAP (1972) *Another Chance for Cities: SNAP 69/72*, Liverpool: Shelter Neighbourhood Action Project sponsored by SHELTER.

Southampton (1991) *Women and the Planned Environment: Design Guidelines*, Southampton: Directorate of Strategy and Development.

Southwark (1989) *Housing Security Design Guide*, Southwark: Planning Department.

Spain, Daphne (1992) *Gendered Spaces*, London: University of North Carolina.

Spallone, Patricia (1989) *Beyond Conception: The new politics of reproduction*, London: Macmillan.

—— and Steinberg, Deborah (1987) *Made to Order: The myth of reproductive and genetic progress*, Oxford: Pergamon.

Spencer, Anne and Podmore, David (1987) *In a Man's World: Essays on women in male-dominated professions*, London: Tavistock.

Spender, Dale (1982) *Women of Ideas: And what men have done to them*, London: Routledge & Kegan Paul.

—— (1983) *Feminist Theories: Three centuries of women's intellectual traditions*, London: Women's Press.

—— (1985) *There's always been a woman's movement*, London: Routledge & Kegan Paul.

—— and Spender, Lynn (1983) *Gatekeeping: Denial, dismissal and distortion of women*, London: Pergamon.

Sprague, Joan (1991) *More than Housing: Lifeboats for women and children*, London: Butterworths.

Squier, Susan, Merrill (ed.) (1984) *Women Writers and the City: Essays in feminist literary criticism*, Knoxville, Tennessee: University of Tennessee Press.

Stacey, Margaret (1960) *Tradition and Change: A study of Banbury*, Oxford: Oxford University Press.

—— and Price, Marion (1981) *Women, Power and Politics*, London: Tavistock.

Stanley, Liz (1987) 'Some notes on "hidden" work in public places: the case of Rochdale', in Liz Stanley, *Essays on Women's Work and Leisure and "Hidden" Work*, Manchester: University of Manchester.

—— (1990) *Feminist Praxis: Research, theory, and epistemology, in feminist sociology*, London: Routledge.

—— and Wise, Sue (1983) '"Back into the Personal" or: our attempt to construct "feminist research"' in Gloria Bowles, and Renate Klein Duelli (eds) *Theories of Women's Studies*, London: Routledge & Kegan Paul.

—— —— (1993) *Breaking out Again: Feminist ontology and epistemology*, London: Routledge.

Stanworth, Michelle (1987) *Reproductive Technologies: Gender, motherhood, and medicine*, Oxford: Polity.

Steinam, Gloria (1992) *Revolution from Within: A book of self-esteem*, London: Bloomsbury.

Stephen, Douglas, Frampton, Kenneth and Carpapetian, Michael (1965) *British Buildings 1960–64*, London: A. and C. Black.

Stewart, Ian and Golubitsky, Martin (1991) *Fearful Geometry: Is God a geometer?*, Oxford: Blackwell.

Stewart, Katie (1981) 'The marriage of capitalist and patriarchal ideologies: Meanings of male bonding and male ranking in U.S. culture', in Lydia Sargent (ed.) (1981) *Women and Revolution*, London: Pluto.

Stimpson, Catherine, Dixler, Elsa, Nelson, Martha and Yatrakis, Kathryn (eds) (1981) *Women and the American City*, Chicago: University of Chicago Press.

Stone, Irving (1961) *The Agony and the Ecstasy: A biographical novel of Michelangelo*, Collins: Fontana.

Stone, John (1983) 'Couples are out of order', *Planning* 546: 2, 25 November.

Strauch, Renate and Wirthwein, Martina (eds) (1989) *Beim Zweiten Blick Wirkt Alles Anders: Frauen dokumentieren ihre Stadt* (At a Second Glance Everything Seems Different: Women write about their city), Swerte, Germany, Women and Planning Group, City of Swerte (via Yorkshire branch of the RTPI).

Strauss, Anselm (ed.) (1968) *The American City*, London: Allen Lane.

Stübben, Joseph (1907) *Handbuch der Architektur*, Vienna (originally published as a series in the architectural journal *Baumeister*, see Collins and Craseman Collins, 1966).

Sturtevant, Katherine (1991) *Our Sister's London: Nineteen feminist walks*, London: Women's Press.

Summerson, John (1978) *Georgian London*, Harmondsworth: Penguin.

Surrey County Council (1965) *Surrey Development Plan: First Review: Report and analysis of survey*, Surrey County Council.

Susskind, L. E. (1984) 'I'd rather invent the future than discover it', *Journal of Planning Education and Research* 3 (2): 89–90.

Sutcliffe, Anthony (1970) *The Autumn of Central Paris: The defeat of town planning 1850–1970*, London: Edward Arnold.

—— (1974) *Multi-Storey Living: the British working class experience*, London: Croom Helm.

Swenarton, Mark (1981) *Homes fit for Heroes*, London: Heinemann.

Swords-Isherwood, Nuala (1985) 'Women in British Engineering', in Wendy Faulkner, and Erik Arnold, (eds) *Smothered by Invention: Technology in women's lives*, London: Pluto.

Tabor, Roger (1991) *Cats: The rise of the cat*, London: BBC Publications.

Tannahill, Reay (1989) *Sex in History*, London: Hamish Hamilton.

Tawney, Richard (1922) *Religion and the Rise of Capitalism*, Harmondsworth: Penguin, 1966.

Taylor, Barbara (1984) *Eve and the New Jerusalem*, London: Virago.

Taylor, Beverley (1985) 'Women plan London', *Women and Environments* 7 (2): 3–7, Spring.

—— (1988) 'Organising for change within local authorities: how to turn ideas into action to benefit women', paper given at *Women and Planning: Where next?*, London: Polytechnic of Central London: Short Course Report, 16 March.

Taylor, Frederick Winslow (1895) *Scientific Management*, London: Harper & Row, 1964.

Taylor, Judith (1990) 'Planning for women in unitary development plans: an analysis of the factors which generate "planning for women" and the form this planning takes', unpublished MA thesis, Sheffield University: Town and Regional Planning Department, September.

Taylor, Nancy (1986) 'GEAR: Are they getting it right?', in Marion Chalmers (ed.) *Women and Planning in Scotland: New communities – did they get it right?*,

Women and Planning Working Party, Royal Town Planning Institute, Scottish Branch, and West Lothian District Council.

Taylor, Nicholas (1973) *The Village in the City*, London: Maurice Temple Smith.

TCPA (Town and Country Planning Association) (1987) 'A Place for Women in Planning', *Town and Country Planning* 56 (10): 279.

—— (1990) 'Special Issue on Racial Issues', *Community Network*, Autumn.

Teather, Elizabeth (1991a) 'Visions and realities: images of early post war Australia', *Transactions*, 16 (4): 470–83.

—— (1991b) 'Planning documents: as value-laden and selective as fiction? The Cumberland County planning scheme, Sydney, 1948', *Planning History: Bulletin of the planning history group*, 13 (2) 1991.

Telling, John (1990) *Planning Law and Procedure*, London: Butterworth.

Tetlow, John and Goss, Anthony (1968) *Homes, Towns and Traffic*, London: Faber.

Thomas, Edith (1966) *The Women Incendiaries*, London: Secker & Warburg.

Thomas, Huw, and Krishnarayan, Vijay (1993) 'Race Equality and Planning', *The Planner* 79 (3): 17–19, March.

Thomas, Keith (1988) *Development Control Distance Learning Package, Issues in Development Control: Unit 4, The Home as Workplace*, Oxford: Oxford Polytechnic in conjunction with the Royal Town Planning Institute.

Thompson, E. P. (1963) *The Making of the English Working Class*, Harmondsworth: Penguin.

Thompson, F. Michael L. (1963) *English Landed Society*, London: Routledge & Kegan Paul.

—— (1968) *Chartered Surveyors: the growth of a profession*, London: Routledge & Kegan Paul.

Thompson, Jane (1983) *Learning Liberation: Women's response to men's education*, London: Croom Helm.

Thurston, Hazel (1974) *Royal Parks for the People*, London: David & Charles.

Tickner, Lisa (1987) *The Spectacle of Women: Imagery of the suffrage campaign 1907–14*, London: Chatto & Windus.

Tönnies, Ferdinand (1955) *Community and Association*, London: Routledge.

Torre, Susana (ed.) (1977) *Women in American Architecture: A historic and contemporary perspective*, New York: Whitney Library of Design.

Trevelyan, G. M. (1978) *English Social History*, Harlow: Longman.

Tripp, Alker (1942) *Town Planning and Traffic*, London: Metropolitan Police Authority.

Tripp, Maggie (ed.) (1974) *Woman in the Year 2000*, New York: Dell.

Tuck, Andrew (1993) 'The lavver's guide', *Time Out*, 24–31 March: 30–2.

Tudor Walters Report (1918) *Report of the Committee on Questions of Building Construction in Connection with the Provision of Dwellings for the Working Classes*, London: HMSO.

UKIPG (1990) *Women in the Professions*, London: The United Kingdom Inter-Professional Group, Working Party on Women's Issues.

Underwood, Jacky (1991) 'What is really "material"? Rising above interest group politics', in H. Thomas, and P. Healey, *Dilemmas of Planning Practice: Ethics, legitimacy, and the validation of Knowledge*, Aldershot: Avebury.

Ungerson, Clare (1985) *Women and Social Policy*, London: Macmillan.

Ussher, Jane (1991) *Women's Madness: Misogyny or mental illness*, Lewes: Harvester.

Uthwatt Report (1942) *Report of the Expert Committee on Compensation and Betterment*, London: HMSO.

Vaiou, Constantina (1990) 'Gender Relations in Urban Development: An alternative framework of analysis in Athens, Greece', Ph.D. thesis, University College London.

—— (1992) 'Gender divisions in urban space: beyond the rigidity of dualistic classifications', *Antipode* 24 (1): 247–62.

Vallance, Elizabeth (1979) *Women in the House*, Athlone Press, London.

Van Veen, Paf and Van Der Sijs, Nicoline (1991) *Etymologisch Woorden Boek: de herkomst van onze woorden*, Utrecht: Van Dale Lexicografie.

Veblen, Thorstein (1971) *Theory of the Leisure Class*, London: Allen & Unwin.

Venables, P. (1955) *Technical Education: Its aims, organisation and future development*, London: G. Bell & Sons.

Voigt, W. (1989) 'The garden city as eugenic utopia', *Planning Perspectives* 4: 295–312.

Vonarburg, Elisabeth (1990) *The Silent City*, London: Women's Press.

Vliet van, Willem (ed.) (1988) *Women, Housing and Community*, Aldershot: Avebury.

WAC (Women Architects Committee) (1993) *Women Architects*, London: RIBA.

Wajcman, Judy (1991) *Feminism Confronts Technology*, Cambridge: Polity.

Walby, Sylvia (1986) *Patriarchy at Work*, Cambridge: Polity.

—— 1990) *Theorising Patriarchy*, Oxford: Blackwell.

Walker, Lynn (1989) 'Women and architecture', in Judy Attfield and Pat Kirkham (eds) *A View from the Interior: Feminism, women, and design*, London: Women's Press.

Walkowitz, Judith (1992) *City of Dreadful Delight: Narratives of sexual danger in late Victorian England*, London: Virago.

Walsh, Linda and Gibson, Mike (1985) 'The average planner and CPD', *The Planner*, 71 (3), March.

Walter, Marianne (1985) An Exile in England, unpublished manuscript.

—— (1992) *The Poison Seed: A personal history of Nazi Germany*, Lewes, Sussex: The Book Guild.

Walters, Derek (1989) *Chinese Geomancy: Dr. J. J. M. De Groot's seminal study of Feng Shui with detailed commentaries*, Shaftesbury, Dorset: Element Books.

Waltham Forest (1988) *Access for All*, Waltham Forest: Planning Department.

Ward, Stephen (1978a) *The Child in the City*, London: Architectural Press.

—— (1978b) 'The House that Jack Built', *BEE: Bulletin of Environmental Education*, Town and Country Planning Association, August–September, 88–9, whole issue.

—— (1987) 'The lady tracers', *Town and Country Planning* 56 (10): 255–6, October.

—— (1991) 'The fairy ring', *Planning History* 13 (1): 29–32.

Ward, Dorcas (1963) 'The work of a housing manager', *The Chartered Surveyor* 95 (7): 392, January.

Ware, Vron (1987) 'Problems with design improvements at home', *Town and Country Planning* 56 (10).

—— (1992) *Beyond the Pale: White women, racism and history*, London: Verso.

Warin, Anne (1989) *Hilda: An Anglo-Saxon chronicle*, London: Lamp Press.

Waring, Marilyn (1988) *If Women Counted: A new feminist economics*, London: Macmillan.

Warner, Marina (1978) *Alone of All Her Sex*, London: Quartet.

Watson, Sophie (1990) *Playing the State: Australian feminist interventions*, London: Verso.

Wayte, Gillian (1986) 'Stories of genius: some notes on the socialisation of fine art students', in D. Farran, S. Scott and L. Stanley, *Writing Feminist Biography*, Studies in Sexual Politics, Manchester University.

—— (1989) 'Becoming an artist: the professional socialisation of art students', unpublished Ph.D., University of Bristol.

WDS (Women's Design Service) (1990 current publications) *It's Not All Swings and Roundabouts; Making better playspace for the under sevens, Women and Safety on Housing Estates; Thinking of Small Children: Access, provision and play; Shoppers'*

Creches: Guidelines for childcare facilities in public places; *At Women's Convenience: A handbook on the design of women's public toilets; Accessible Offices*, London: Women's Design Service.

—— (1992) *Challenging Women: City challenge*, London Women's Design Service, occasional newsheet, Autumn.

—— (1993) *Race and Gender in Architectural Education*, Broadsheet, Women's Design Service, The Print House, 18 Ashwin Street, London E8 3DL (071 241 6910) and Society of Black Architects, 30 Chalfont Court, Baker Street, London NW1 5RS.

WEB (Women and the Built Environment) (1990) 'Simplistic, inadequate, and tokenistic', Comment in *Newsletter of Women in the Built Environment* 17: 3.

Webb, Sue (1990) 'Counter-arguments: an ethnographic look at "women and class"', in Liz Stanley (ed.) *Feminist Praxis: Research, theory and epistemology in feminist sociology*, London: Routledge.

Weber, Max (1946) *The Methodology of the Social Sciences*, New York: Free Press.

—— (1964) (introduction by Talcott Parsons, (ed.)) *The Theory of Social and Economic Organisation (Wirtschaft und Gesellschaft)*, New York: Free Press.

Webster, B. (1983) 'Women's committees', *Local Government Policy Making* 10 (2).

Weiner, Gaby (ed.) (1985) *Just a Bunch of Girls*, Milton Keynes: Open University.

Weisman, Leslie and Birkby, Noel (1983) 'The women's school of planning and architecture', in Charlotte Bunch, and Sandra Pollock, *Learning our Way*, New York: Crossing Press.

Wekerle, Gerda, Peterson, Rebecca and Morley, David (eds) (1980) *New Space for Women*, Boulder, Colo.: Westview Press.

Wells, H. G. (1933) *Experiment in Autobiography*, New Haven, Conn.: Yale University Press, 1986.

Westcott, Anthony (1992) *How will Europe affect you?*, West Midlands Study Centre and Bristol Polytechnic.

Westcott, Marcia (1981) 'Women's studies as a strategy for change: between criticism and vision', in Gloria Bowles and Renate Duelli Klein (eds) (1983) *Theories of Women's Studies*, London: Routledge & Kegan Paul.

Westergaard, John and Resler, Henrietta (1978) *Class in Capitalist Society*, Harmondsworth: Penguin.

WGPW (Women's Group on Public Welfare, Hygiene Committee) (1943) *Our Towns: A close up*, Oxford: Oxford University Press, with London: NCSS.

WGSG, (1984) (Women and Geography Study Group, Institute of British Geographers) *Geography and Gender*, London: Hutchinson.

Whatmore, Sarah (1991) *Farming Women: gender, work, and family enterprise*, London: Macmillan.

Whitaker, Ben and Browne, Kenneth (1971) *Parks for People*, London: Seeley.

Whitburn, Julia (1976) *People in Polytechnics*, London: SRHE.

Whitelegg, Elizabeth, Arnot, Madeleine, Bartels, Else, Beechey, Veronica, Birke, Lynda, Himmelweit, Susan, Leonard, Diana, Ruehl, Sonja and Speakman, Mary Anne (eds) (1982) *The Changing Experience of Women*, Oxford: Basil Blackwell with the Open University.

Whylde, June (1983) *Sexism in the Secondary Curriculum*, London: Harper & Row.

Whyte, Judith, Deem, Rosemary, Kant, Lesley and Cruickshank, Maureen (eds) (1985) *Girl Friendly Schooling*, London: Methuen.

Whyte, William (1963) *The Organisation Man*, Harmondsworth: Penguin.

—— (1981) *Street Corner Society*, Chicago: University of Chicago Press.

Wigfall, Valerie (1980) 'Architecture', in Rosalie Silverstone and Audrey Ward Audrey (1980) *Careers of Professional Women*, London: Croom Helm.

Wigley, Mark (1992) 'Untitled; the housing of gender', in Beatriz Colomina (ed.) *Sexuality and Space*, New York: Princeton Architectural Press.

Wigram, George (1980) *The Englishman's Hebrew and Chaldee Concordance of the Old Testament: Numerically coded to Strong's exhaustive concordance*, Nashville, Tenn.: Broadman Press.

Wikan, Unni (1980) *Life Among the Poor in Cairo*, London: Tavistock.

Willis, W. A. (1910) *Housing and Town Planning in Britain*, London: Butterworths.

Wilner, Daniel and Walkley, Rosabelle Price (1955) *Human Relations in Inter-Racial Housing*, Minneapolis: University of Minnesota Press.

Wilson, Des (1970) *I Know it was the Place's Fault*, London: Oliphants.

Wilson, Elizabeth (1980) *Only Half Way to Paradise*, London: Tavistock.

—— (1991) *The Sphinx in the City: Urban Life, the Control of Disorder and Women*, London: Virago.

Wiltshire, Stephen (1989) *Cities*, London: J. M. Dent & Sons.

Wirth, Louis (1938) 'Urbanism as a way of life', in P. Hatt and A. Reiss (1956) *Cities and Society: The revised reader in urban sociology*, New York: Free Press.

Witherington, Ben (1988) *Women in the Earliest Churches*, Cambridge: Cambridge University Press.

Wollstonecraft, Mary (1792) *Vindication of the Rights of Woman*, Harmondsworth: Penguin, 1975.

Women and Planning Group: Dublin (1990) *Submission from the Women and Planning Group for the Review of Dublin County Council Development Plan*, Dublin: Irish Planning Institute, and RTPI, Irish Branch, Southern Section.

Women and Planning: Scotland (1990) *Network; Women and Planning in Scotland*, various, e.g. 'Shopping centre, survey results' 'ed Sheila Anderson, no date, and 'Childcare, what is your employer doing?', Issue 14, Nov.1990.

Woods, Mike and Whitehead, Jaccqui (1993) *Working Alone: Surviving and Thriving*, London: Pitmans in association with the Suzy Lamplugh Trust.

Woolf, Virginia (1929) *A Room of One's Own*, London: Collins, 1977.

Wright, Myles H. (1948) *The Planner's Notebook: A compendium of information on town and country planning and related subjects*, London: The Architectural Press.

Wright, Olin Erik (1985) *Classes*, London: Verso.

Young, Iris (1990) 'The ideal of community and the politics of difference', in Linda Nicholson, *Feminism/Postmodernism*, London: Routledge.

Young, Michael (1958)*The Rise of Meritocracy*, Harmondsworth: Penguin.

Young, Michael and Willmott, Peter (1957) *Family and Kinship in East London*, Harmondsworth: Penguin.

—— and —— (1978) *Symmetrical Family: Study of work and leisure in the London region*, Harmondsworth: Penguin.

Zmroczek, Christine (1992) 'Dirty linen: women, class and washing machines, 1920s–1960s', *Women's Studies International Forum* 15 (2): 173–86.

NAME INDEX

Abercrombie, Patrick 117, 132, 149
Abrams, Mrs 108
Adams, Thomas 101, 110
Adburgham, Miss 103, 108
Adshead, Stanley 111, 117, 126, 133
   (footpath width), 183
Aiton, Norah 108
Albery, Jessica 135
Albinski, Nan Bowman 97
Alcott, Louisa May 97
Allen, Professor 112
Amooquaye, Eno 46
Amos: Geraldine 167; Jim 167, 146
Anne, Queen 86
Aristotle 117
Armstrong, Jennifer 64
Artemis 80 (unnamed)
Ashworth, Graham 77, 88, 107, 148
Aspinal, Tony 48
Attfield, Judy 46
Augustine 84
Austen, Jane 97
Austin, Alice Constantine 101; Winny, 1

Barclay, Irene 108
Barker Benfield, Ben 105, 116
Barnes, Deborah 115, 168, 170, 186
Barnett, Henrietta 3, 97, 107, 109, 113,
   165
Barrett, Susan 167
Barson, Mary 108
Battersby, Christine 1, 66
Bauer, Catherine Wurster 144
Beauvoir, Simone de 13
Becker, Lydia 96
Bedford, Dowager Duchess 86
Beethoven 116
Bell, Colin and Rose 92

Bellamy, Edward 98
Benfield, Eileen 108
Bennington, John 170
Benson, Betty 109
Benzerfa-Guerroudj 62
Besant, Annie 95, 99
Beveridge 125
Bhride, Karla Ni 44
Biswas, Aloke 149
Blackett, Colette 170
Blavatsky, Madame 99
Blomfield, Margaret 109
Bloomsbury Group 86, 99
Bondfield, Margeret 113
Booth, Charles 99
Booth, Christine 170
Booth, William 99
Boulding, Elise 81
Boys, Jos 170, 175
Brand, Janet 64, 162
Bristol Women's Architects Group 168
Brooke, Dorothea 97
Brooke, Jane 167
Brown, Denise, Scott 168 see also
   Venturi
Brueghal, Irene 128
Buchanan, Colin 134
Bulos, Marjorie 175
Bunyan, John 85
Burdett-Coutts, Angela 95, 97
Burgess, Jacqueline 181

Cadbury, George 94–5, 99, 112
Cadbury, Ruth 170, 175
Calvin 85
Carr, Jonathan 96
Carson, Rachel 63
Cashmore, Hilda 108

Castells, Manuel 152
Castle, Barbara 135
Castledine, Pat 167
Catran, Margaret 166
Cavanagh, Sue 43, 175
Chadwick, Edwin 91
Chapin, F. Stuart 142
Chapman, Honor 165
Charles, Ethel 108
Cherry, Gordon 73, 88, 101, 107, 145
Chesterton, Elizabeth 108
Christaller, Walter 126
CISC 19, 67
Clough Williams, Ellis 110, 145
Clutton, Cecil 135
Cockburn, Cynthia 55, 135, 159, 165
Cole, Annette 108
Cole, Margaret 109
Coleman, Alice 64
Coleman, Hilary 64
Colenutt, Bob 22
Comfort, Doctor Alex 105
Corbusier, Le 1, 83, 121–2
Cordingley, Professor 117
Creese 136
Crofts, Angela 168
Crowe, Sylvia 135

Dale, David 92
Dale's daughter 92
Dan the Plan 154
Darke, Jane 168, 175; John 168
Darley, Gillian 71, 84, 96–7
Darwin, Charles 116
Davies, Anne 167
Dean, John 171
de Beauvoir, Simone 13
de Pizan, Christine 84
Deney, Diane de 166
Delamont, Sara 14, 38, 142
Denby, Elizabeth 3, 108, 138, 145
Denington, Evelyn 3, 114
Denning, Lord 145
Dent, Mary 20
de Sade 104
Descartes 117
Despard, Charlotte 113
Doxiadis 117, 181
Dresser, Madge 66
Drew, Jane 109, 122, 168
Dumphy, Norah 108
Duncalfe, Margot 171
Duncan, Carmen 179

Duony, Andreas 168 *see also* Zyberg

Eanaigh, Cristin Ni 190
Eliade 11–12, 53, 70
Eliot, George 97
Ellis, Clough Williams 110, 145
Ellis, Havelock 99
Engels 97
Essex, Sue 33
Evans, Bernard 161
Eversley, David 127

'faded lady' 165
Fagg, Christopher 116
Fairbrother, Nan 64
Farrish, Maureen 175
Fawcett, Millicent 123
Fidler, Peter 161
Fitzsimmonds, Diana 167
Flintstones 75
Foster, Wendy 168; Norman 168
Foucault, Michel 189
Foulsham, Jane 4, 166, 170
Fourier, 93, 105, 121
Frankenstein 97
Freeman, Roger 4
French, Marilyn 80
Freud 52, 77, 116, 152
Friedan, Betty 8, 41
Fyson, Anthony 181

Gale, Andrew 25
Gandhi, Sonia 95
Gans 41
Gavron, Hannah 146
Geddes, Patrick 80, 90, 116–17, 123, 149
Gibbon, Frederick 158
Gibson, Mike 159
Gilligan, Carol 54
Gilman, Charlotte Perkins 98, 101
Gilson, Caroline 108
Ginsberg, Leslie 137
Glass, Ruth 3, 143–4
Goodison, Lucy 74
Goring, Anne 166
Goss, Professor 147
Greenaway, Kate 96
Grey, Eileen 122
Gudjonssen, Susan 170

Hall, Peter 90–1
Hanson, Caroline 181
Haraway, Donna 14, 52, 54

Harran, Betty 27
Harvard Square group 101
Harvey, David 9
Hass-Klau, Carmen 187
Havelock Ellis 99
Hayden, Dolores 7, 85, 94, 97, 179
Healey, Patsy 9, 21, 87, 152
Heap, Desmond 112
Hegel 38
Henniker-Heaton 158
Hilda, St 84
Hill, Octavia 99–100, 111, 114
Hillel, Mira Bar 166
Hillier, Jean 48
Hillman, Judy 166
Hincliffe, Tanis, 86
Hooper, Gloria 166
Howard, Ebenezer 1, 31, 93–4, 99, 110;
   Mrs E. 108
Howatt, Hilary 162, 167
Howe, Elizabeth 10
Hoyle, Christine 165
Hoyt, Homer 115
Hutchinson, Max 30
Huxley, Aldous 116

Iroquois 123

Jackson, June 170
Jackson, Michael 79
Jacobs, Jane 3, 15, 49, 163
James, Carol 166
James, Professor J.R. 148
Jamieson, Jill 168
Jarrett, Valerie 172
Jeffreys, Sheila 104
Jeffries, Leonard 79
Jekyll, Gertrude 97, 99
Jenkins, Leila 109
Jones, Caren Prys 16
Jones, Sally 3, 26, 159
Joseph, Martin 9

Kanter, Rosabeth Moss 105
Keeble, Lewis 34, 147
Keller, Suzanne 150
Kershaw, Diana 171, 175
Kimball-Hubbard, Theodore 109
King, Graham 75
Kingsley Kent, Susan 117
Kinnock, Neil 170
Kirk, Gwyneth 19
Kitchen, Ted 158, 165

Kumar 104

Ladies Lavatory Company 104
La Union de Femmes 122
Lambert, Lydia 168
Lamplugh 177
Law, Sylvia 1, 159, 165
Le Corbusier 1, 83, 121–2
Le Play 117
Lee, Anne 106
Lees, Audrey 155, 165
Leighton Lockton, Sarah 167
Lever, William 95, 99, 111
Lever Brothers 1
Lichfield, Dahlia 168
Little, J 4, 39
Livingstone, Ken 168
Loftman, Patrick 134
Lom, Wivi 122
Louis XIV 85
Lutyens, Edwin 95; Lutyens' daughter
   91
Lydia, seller of purple 83

McBride, Catherine 175
McCaffrey, Ann 98
McCall, Cicely 113,
McCallum, Ian 118, 121
McCrae, Heather 133
McDowell, Linda 11, 22, 36, 52, 154, 188
McKay, Hazel 1, 167
Mackeith, Margaret 103, 132
Mackenzie, Suzanne 3, 59, 102
McLoughlin, J. 139, 177
Madame Blavatsky 99
Malthus 60
Mann, Jean 125
Marcus, Susanna 109, 159, 161
Marie Antoinette 97
Marx: Eleanor 99; Karl 37, 90, 97
Massey, Doreen 8
MATRIX 16, 48, 63, 119, 175
Mead, Margaret 7, 146
Mendis, Sunethra 171
Merrett, Tracy 168
Michelangelo 68
Minett, Miss 159
Mitchell, Elizabeth 146
mohawk 123
Moneypenny, Joan 113
Moore, Victor 24
More, Thomas 84
Morgan, Elaine 73

Morgan, Garrett 134
Morgan, Julia 108
Morphet, Janice 164
Morris, Terence 2, 143
Morrison, Herbert 114, 147
Mumford, Lewis 75, 117

Nadin, Vincent 26, 159
Nash, John 86
Newby, John 167
Nin, Anaïs 99, 191
Nott, Susan 24, 184
Nubians 78

Osborn, Frederick 110, 117
Owen, Alex 100
Owen, Robert 92, 111

Pahl, Ray 56, 151
Papafio, Kwarley 176
Parker, Barry 95, 116, 119
Parker Morris 47
Parkes, Rosa 149
Parkin, Frank 25
Pateman, Carol 38
Payne, Gillian 27
Pearsall, Phyllis 109
Pearson, Lynn 7
Peirce, Melisia Fay 101
Pepler, Lady 113
Pepler, George 117
Phillips, Yvonne 133
Piercy, Marge 98
Pitt, David 165
Pizan, Christine de 84
Plato 30, 79, 121, 152
Polyanna 154
Pollard, Wendy 168
Pomeroy, Florence 96
Pound, Wendy 27
Priestley, J.B. 2, 120
Prince Charles 66, 122, 123, 129, 157
Prince of Wales see Prince Charles
Princess Noor of Jordon 157
Purdom 94
Purvis, June 109, 113
Pythagorus 83

Queen Anne 86

Rachewsky, Dina 172
Rakodi, Carol 61
Rao, N. 176

Rathbone, Eleanor 113
Ravetz, Alison 71, 113, 138
Reade, Eric 18, 55, 146
Reeves, Dory 170
Rex and Moore 148
Reynolds, Josephine 159
Richards, Ellen Swallow 91, 101
Roberts, Marion 1, 90, 116, 122
Roberts, Margaret 45, 167
Roberts, Patricia 175
Rodgers, Richard 122
Rodriguez-Bachiller, Augustin 21–2
Rogers, Barbara 21, 40, 44, 52
Rogers, Elsie 144
Rosenau, Helen 71
Rossetti, Florence 150
Ruff, Alison 166
Rydin, Yvonne 19

Salt, Titus 93
Sant' Elia 123
Satchell, Anne 166
Savill, Maria 85
Schoffield 132
Schreiner, Olive 99
Scott, Elizabeth 3, 122; Betty 108 see also Aiton
Seale, Annette, 132
Serota, Baroness 114
Shakers 106
Sharp, Evelyn 3, 114
Sharp, Thomas 116
Shaw, George Bernard 90, 99
Shaw, Norman 96
Shelley, Mary 97
Shepley, Chris 64
Shoard, Marion 64
Silkin, Lewis 145
Simmie, James 144
Sitte, Camillo 116, 149
Smith, Muriel 138
Smith, T. Dan 147
Somerville, Una 162
Spallone, Patricia 60
Spender, Dale 88
Stacey, Margaret 141, 145
St Hilda 84
Stjernstedt, Rosemary 108
Stopes, Marie 99
Stowe, Catherine, Beecher 101
Stübben, Joseph 149
Summerson, John 86
Superman 30

Susskind, L.E. 54
Sutcliffe, Anthony 86
'Sylvia, mysterious' 154

Takmaz, Sule 170
Tassell, Janet 167
Taylor, Beverley 166, 168
Taylor, Frederich 119
Taylor, Judith 168
Taylor, Maureen 148
Taylor, Nicholas 150
Teather, Elizabeth 2
Tereshkova, Valentina 58
Tertullian 84
Thompson, F.L.M. 151
Thompson, J. Arthur 116
Thompson, Robin 166
Tibbalds, Francis 68
Tönnies, Ferdinand 38
Toynbee, Arnold 99
Tripp, Alker 133
Tudor Walters 119

Underwood, Jackie 33, 166
Unwin, Raymond 95, 116, 118; Lady
    125 (obituary)

Vaiou, Dina, Constantina 79
Venturi, Robert 168
Vonarburg, Elisabeth 98

Wagner, Otto 149 (planning); Richard
    116 (opera)
Wajcman, Judy 42
Walby, Sylvia 38
Walker, Lynn 108
Walkely, Rosabelle Price 149
Walsh, Linda 159

Walter, Marianne 132
Ward, Colin 47
Ware, Vron 175
Waterford, Marchioness of 96
Watson, Sophie 6, 163
Wayte, Gillian 68
WDS 16, 43, 63, 119, 175 see also
    MATRIX
Webbs 99
Webb, Beatrice 100, 113
Weber, Max 10, 13, 31, 153
'weeping girls' 149
Wells, H.G. 105
White, Elizabeth 158
'wildmen' 148
William the Conqueror 85
Willis, Margaret 165
Willmott, Peter 141
Wilson, Elizabeth 72, 102, 125
Wilson, Harold 148
Wiltshire, Stephen 67
WIP (Women in Property) 168
Wise, Valerie 169
Wollstonecraft, Mary 97
Woolf, Virginia 100
Wright, Frank Lloyd 123
Wright, H. Myles 130, 134
Wurster, Catherine Bauer 144

Young, Ivy 1
Young, Lady 114
Young, Michael 141, 144
Yule, Betty 223

Zephaniah 77,
Zmorczek, Christine 73
Zyberg, Elizabeth Plater 168 see also
    Duony

SUBJECT AND PLACE INDEX

'A–Z' 109
abbesses 84
aboriginees 74
abortion 59
academic women 32, 160
Access for Disabled 185
ad hoc: bodies 27; conditions 186
adolescent boys 151
Adullam, cave of 143
'adult' women 148
'affluent worker' 142
Africa 52, 63, 78, 149 see also black, ethnic minorities
Afro Caribbean hairdressers 175
agriculture 57, 62, 145, 182
agorophobia 43, 80 see also platzangast
'All Mod Cons' 43
amazonian age 74
American: Institute of Planners 101; Planning Association 166; suburbs 43
'angel' or 'whore' 80
animals 34, 62, 65, 181–2, 187
apartheid 149
apparatchiks 21
archeology 74
architects 47, 66–9, 85–7
Arndale shopping centre 132
art: 'A' level 68; college 1; artist 124; artist planner 66; arts and crafts 67, 96; art and design 108
artisan cottage 119
Asian women's centre 62–3, 176
astronaut 58
Australia 2, 61, 48
autistic 67
Automobile Association 133, 178
autopia 123

B1 Business Use Class 180
baby: boom 2, 116; buggy 48, see chs 3, 10, 11; changing areas 190; sprawling 144
balance, urban 41
baptism 176
Basildon New Town 146
Bath 86
battery cars 153
Bauhaus 122
beaux-arts 55, 66
Bedford Park 96
Beeching cuts 42
belief 11–13
Benin urban culture 78
Bethnall Green 141
betterment levy 184–6 see also planning gain
betting offices 183
bicycles 44, 94
bingo 39
biography 68
biological division 86
biologist 117
Biotech Global Takeover 64
Birmingham: city planning 112, 171, 177: Draft UDP 182; University 94
birth control 59
black: child 67; American engineer 133; men 26; church 150, 176; people 43; prophet 77; women 25–6, 175
blind corners 48
Bloomsbury 86, 91, 99
Blott on the Landscape 31
boundary: gypsies 82; railings 178; religious 78
bourgeois feminist 82, 102
Bourneville 94

bowler hats 148
Bradford Halifax 93
brat pack 138, 146, 148
breast feeding 104
'bricks without straw' 78
Bristol 150, 171, 175; SAUS 114, 166;
    Women's Advisory Housing
    Committee 108; Women's Architects
    Group 168; Women in Property 168
British Rail 178 see also train
Brixton, South London 150
BS (British Standard): 5750 5; 5810 185
building societies 93
bungalows 2, 121
'burbs' 41 (hoods 46)
bus stops 49
business: cultures 187; woman 58
buttocks 75

CAD (Computer Aided Design) 69
Cadbury Trust 171
cafés 131, 179
Cairo 78
calculus 157
Calvin's Geneva 85
Camden Hopper bus 185
cameo photograph 107
Canada 131
Canterbury cathedral 115
capital receipts 185
car 42, 53; company 44; parking 134,
    176, 178
careers in town planning 32–3
cartoon: of a little boy 167; bowler hats
    148; Dan the Plan 154; Polyanna 154
caryatides 83
cattle pens 138
cats 2, 47
celtic women 84
central area office occupants 44–5
centralised food production 102
chameleons: planners 110; women 163
Chandigargh 122
Changing Places 4, 174
Cheshire 148
Chicago School of Sociology 115, 142
child: care 24, 48, 101, 150; compound
    131; minders 42, 179; play, 48; as a
    minority 34
chair 183
Children Act 1989 48, 179
china 59
CIS 57–8 see also Soviet Union

cholera 91
CISC (Construction Industry Standing
    Conference) 19, 66
citizenship 38
city state 79
Civic pride 103
civil engineering 157
class 25, 37; love/hate 40; middle 120;
    working 91 see also housing
'clean up the city' 62
Clifton 150
closure 25
cloth caps 128
CND 27
co-operative housekeeping 94, 101, 119,
    179
Code of Conduct 172, 185, 191, 193
College of Estate Management 157
colonialism 8, 149
communitarianism 89, 94, 97
community 46, ch. 9, 142, 150
commuting 126, 133
company car 44
competence (NVQ) 67
computers 53, 139
conditions in planning permissions
    184–6
construction industry 19
consumption 125, 127–31
Contagious Diseases Acts 91
container, planning as 7
contraception 59
conveyor belts 102
corrective zoning
cost benefit analysis 135
council housing 91, 109, 118
councillors 5, 186; feminist 5
countryside 145, 181 see also
    agriculture, open space
'couples out of order' 168
Coventry 109, 131, 172
craft, 67, 122 see also art
Crawley 127
creativity 67 see also genius
creche 4, 7, 129, 175, 185
'creche course for developers' 183
credentialisation 156
criminal: activity 143; area 2
credibility gap 55, 70
'crit session' 112
cul-de-sac 133
cusp 116 see also spinode
cybernetics 139

cyborg 64

dance centres 176
dead city centres 178
decentralisation 115 *see also* density
Deeplish 171
defensible space 151
delusions of grandeur 19
density: debate 115–16, 138–9, 144; FSI
    129
Deptford 176–7
detached house 121
determinism: environmental 145, 151;
    geographical 79; historical 73–4
development; definition 24; control,
    23–4; feminism and 34; gender and
    34; men and 51; plan 6, 125, 140, 170;
    planning 32; women and 39
devil's gateway 84
dezoning 178
dichotomies 12
disability 153, 174; resource team 176;
    disabled 26, 47, 167, 169, 185
disasters 53
dispensationalist theory 74
dissenter and puritan sects 85
distance learning unit 111, 180
dogs 34, 47, 181–2
domestic science 91
Doncaster 171, 177
draughtpersons 132
driving lessons 43
Dronfield Pioneer Housing, Sheffield
    132
dropped kerbs 4, 153
Dublin 181
duck ponds 137, 181
Durham 112
DWEBS 78

Ealing 178
East End 141
East Kilbride 146
economists 23
Edinburgh 87, 112, 123
Egypt 78
elderly 146; women 39
Electrical Association for Women 113
Electricity Women's Group 113
Elephant and Castle 131
Enlightenment 52
enterprise culture 164
entrepreneurial woman 82

enveloping 174
environmental: areas 134; assessment
    65; determinism 145, 151; quality 65
EO (Equal Opportunities) 5, 65, 79,
    162, 169
EQ (Environmental Quality) 65
erotic meanings 83, 89, 104–5, 116
eruv 137
Esperanto 95
'essential' users 44
Essex Design Guide 48
estate design 47–9, 151
Estate outside the Dwelling 151
Ethiopia 79
ethnic minorities 3, 39, 60, 66, 153, 159
ethnographic: anecdotes 13; eugenics
    8, 59, 89–90, 101, 115; studies 142;
    Review 60; Quarterly 60
Euro Disney 70
Europe 18, 46, 121
evangelicalism 89, 99–100
Evangelistic Bureaucrat 146
evening use of shopping centres 183
evolution 117
exemplar UDP policy statements ch. 11
    (and LWPG in bibliography)

Fabians 99
factories 46, 117, 127
family: man 139; planning 59
Fawcett Society 123
feminism: definition 34; first wave 89;
    and throughout
femocrat 6, 82, 163
feng shui 72, 78
Festival of Britain 143
field 107, 112, 137
'figures in the landscape' 66
'filching fags' 158
'filthy city' 77
'fine strapping wenches' 126
first undergraduate degree, 1946 156
'fit' 61, 116 *see also* eugenics
flat footed planners 132
flaveuse 102
flexitime 27
folkloric, superstructure 56
food stamps 43
football 48, 139
footbinding 43
footbridge improvement 178
footpaths 62, 133; and shared with
    cycleways 44

'force evolves form' 106, 115
'form follows function' 106, 122
free love 104
free thinkers 99
French urban feminists 46, 122
Friends of the Earth 64
Frontierland 52
FSI (floor space index) 129
fur coats 65
future, creating it 54, 139, ch. 11
futurism 123

garden 47, 97, 137; city 93, ch, 6–7;
    front 2, 'planning merely gardening'
    138
garrison towns 83
gas association 113
gatekeepers 16, 25
GEAR, Glasgow 150, 167
Gemeinschaft, Gesellschaft 38
gender city 186
genetic engineering 60
genius 1, 30, 66, 123
geographers 23, 126, 147, 161, 167
Geographia A–Z maps 109
Geographical Informations Systems
    69
'geography and gender' 4, 188
geomancy 71 see also feng shui
Georgian 85
'getting by' 164
girls: job opportunities 158;
    recreation ground 95
GIS (Geographical Information
    Systems) 69
glass ceiling 26
GLC 4, 6, 169–70; chairman 165;
    women's committee 169
GLDP 132, 169
GLEB 130
gnostics 53
'go into a club' 50
goddesses 74
Golden Rule 83
'gospel of planning' 146
graffiti 122
grammar school boys 157
Grand: Manner 85; Tour of Europe 67
grandes ensembles 68
grass roots planning 144
Greek: city 79; columns 83; squares 79
green: belt 93, 137; grass 137; planning
    63; village green 96, 136

Greenwich 179
Grotton Saga 154
gypsy 82

Hackney 178, 186
half way to paradise 125
hall of mirrors 189
Hammersmith and Fulham 185
Hampstead Garden Suburb 3, 96
Haringey 170, 177, 186
harassment 50, 155
Harvard 109; Square group 101
healers 57
health facilities 44
heaviness (of industry) 126
helmsman 139
'helper' 29
Henniker-Heaton 158
Herland 98
hero 30, 118
high rise 121, 123, 138 see also density
highways 43, 177
Hillingdon 174, 177, 183
historia 72
historicism 73
Holly Village, Highgate 96
'home to work' 42
home, work, and play 181
home working 180
homo-erotic male 89
Honor Oak 178
horizontal distribution (employment) 28
hot food takeaways 66
Hounslow 176
housing: improvement 150, 174; layout
    47; manager 108, 11, 139; private 120;
    tenure 41; and town planning acts
    109; working classes 91
housekeeping, individual 119
Houses of Parliament 147
housewives 76, 86, 91, 127, 146, 152
human soul 12
husbands 167–8
hygiene 90

ideological: crisis 146; problem of 152
imagineers 70
incest 91
India 95
indirect paths 48
inglenooks 119
insides/outsides debate 12, 47–8, 118,
    179

Institute: Planning 108 *see also* RTPI, and Town Planning); British Geographers 4; Housing 111, 139
inter-racial housing 149
'interesting' townscape 48
intuis 117; intuition 67
Ireland 96, 162, 171, 181
Islington 185
Italy 123

Jagonari 62
jars 62
journalists, women 163, 166
'journey to work' 42
Joy of Sex 105

kabalistic geometry 8, 72
Kensington and Chelsea 20, 183
kitchenless houses 102
kitchens 59, 86, 104, 180

La Union des Femmes 122
Labour: party 114, 125; saving devices 146
Ladies: Association for the Diffusion of Sanitary Knowledge 104; Automobile Club 133; Lavatory Company 104
laissez-faire 120
Lambeth 177, 185
landscape 97, 133, 135; of ideas 83
'latch key kid' 144
lawyers 24, 160, 184–6
LCC 119; housing comittee 114 *see also* GLC and Morrison
Leeds 112
'left wing men/right wing men' 114
Leicester 171
'leisure' trips 44
lesbian 169–70
Letchworth 94, 101
Letraset 135
Levittown 41
Lewisham 177, 183
life experience 14
lighting 43, 49, 103
linear development 116
lino 144
Little Women 97
Liverpool University 2, 111
livestock 134
Llano del Rio 101
Lloyds Bank Building 122

local: facilities 44; government 26, 165, 185
log jam 26
London 26; boroughs 20, 170; County Council 114; Londoner and his family 132 *see also* GLC, LCC, GLEB
Long Island 43
Looking Backward 98
Los Angeles 115
Lothian, Scotland 177
low noise level 176
L'Unité d'Habitation 121
Lyons Cornerhouse 103

Mabinogion 98
'machine for living in' 122
maggots 130
male: employment 126; feminist 164; inner city as 142
man: in the street 117; on the 38 bus 134; the hunter 76
managerialism 153
Manchester 112, 177; Disability Study 153
'man, work, place' 117
managerialism 89, 153
mandala 75
manorial hall 119
Mansfield Park 97
marriage 112
'mass production spirit' 121
masters degree 161
mat (mother) 59
matchmakers 57
maternal environment 60
maternity leave 113, 169
mathematical: modelling 135; skills 68; women 166–7
matriarchy 78
maverick 111
maypoles 136
MCD 111
medicine 60
mediterranean culture 82
'men (including women, as throughout)' 158 *see also* boy
mental map 83
meritocracy 146
Merton 177
middle class 76, 60, 147
Middle East 62, 73
Middlemarch 97

migrash 137
military 113, 132, 137
Millicent Fawcett Hall 123
Milton Keynes 44
minibus 176
mining 29, 126, 157
Ministry of Housing and Local
    Government 165
MIT 101, 156
mixed uses 127
mobility 174 *see also* access,
    disability
modularisation 160
moral grounds 100
Morrisonian tradition 114, 147, 169 *see
    also* Labour
Moslem households 180
mosques 176
mother nature 53
motorcar industry 42
MTV 137
multi-nucleated city 184, 187
municipalisation 89

NALGO 170 (now called Unison)
National Playing Fields Association
    136, 176
nature reserves 63
neo-Marxism 56, 89, 147
nepotism 168
netball 136
New Dehli 95
New Lanark 92
new left 151
new technology 121
new towns 46, 48, 110, 127, 145;
    development coroporation 114
New York 123
Newcastle 112, 145, 147, 171
Newham 169, 175, 177, 185
night shifts 48
non-joiners 159
Normans 84
North Vietnam 58
North West of England 126
'nothing gained by overcrowding' 116
Nottingham 112
nurseries 92, 102, 119, 167
NVQ (National Vocational
    Qualifications) 66
offices 27–8, 45, 108, 180, ch. 11
oikos 80
Oneida 105

open plan 119
open space 48, 95, 135–6, 181
opportunity 102
orthodox Jewish families, 180
'other' 14, 52, 54, 74
out-of-town: business parks 49;
    shopping centres 131
outdoor living room 131
outer space 52, 98
overpasses 43
overpopulation 53, 90 *see also*
    eugenics
overspill 115, 127
Oxfordshire 167

paddling pool 181
PAN (planning advice note) 170
Paris: Commune 122
park and ride 44
parks 181
parlour, front 119
Parthenon 80
participation: fraternal 144; public 175
partnership schemes 128
patriarchy 8, 35
'patterns of the past' 87
paving stones 48
pedestrian: man 134; Radburn 133;
    streets 178
pens in a cattle market (gardens) 138
pensioners 170
phalanx community 104
phallic development 116, 123
PhDs 20
philosopher king 30–2, 51, 75, 81, 90;
    queen 81, 160
photographs 155
Picture Post 116
'pierced walls' 178
Pilgrim's Progress 77
pillarisation 149
planning gain 7, 17n4, 185
Plan that Pleased 146
Planners' Realm 154
planning 56; agreement 185; atheism
    146; courses 160; definition 32;
    education 107, 158; gain 185;
    inspectors 27, 145, 193; law 23, 117;
    office 31, 45; RICS division 20
Planning for Choice and Opportunity 4,
    166
platzangst 80
playing fields 137 *see also* open space

plumbing 32, 152
Plymouth 131
police commissioner 133
polluted pump 91
Polyanna 54
poor law guardian, woman 113
Poplar 143
population 40, 59, 61, 116
pornography 59
Port Meirion 110
Port Sunlight 1, 95
Portland Place 147
post graduate degrees 161
post-modernism 122, 153, 188
post-war reconstruction 125
power, eroticisation of 105
practicality 31
prams 132–3 *see also* buggy 48, push
  chair, and chs 3, 10, 11
praxis 151
pre-: industrial cities 38; revolutionary
  Russia 58; school childcare 151
precincts 133–4
predatory professional classes 63
prettifying role 135
priests 74, 84
prior professionals 109
primary industry 126
Princeton 157
private: consultant 133; sector 164
pro-natalist 116
professional: 'front' 19; socialisation
  110, 164; -ism 107
property: above shops 178
prostitution 50, 62, 80–1, 91
prosumption 188
protected category 60
public: conveniences 43, 79, 103–4, 131,
  174, 178, 183 (*see also* 'All Mod
  Cons'); food distribution 112; health
  41, 91, 103; participation 21, 63, 140,
  144–5, 148; squares 80; transport 42,
  62, 103 *see also* open space
pubs 49
Punjab 122
pupillage system 157
purdah 80, 122
push chair 43, 48
Pythagorus, Order of 83

QA (Quality Assurance) 65
Quakers 59
quantitative 52, 138 *see also*

mathematical
Queen Anne 86
queuing 59

R *v.* Donovan 145
R *v.* Boyea 145
race 149; Relations Act 24, 185;
  integration 149
Radburn neighbourhood 133
radiant heaters 131
railings 88
rain forest 63
ramps 153, 178 *see also* steps
rape 75; case 145; of the earth 64
ratecapping 185
rationing 58
'reasonable walking distance' (250 yds)
  131; 177
reconciling conflicting demands 33
recycling 64
Regency 86
Regent Street Polytechnic (PCL,
  University of Westminster) 112
religion 11, 53
repetition 4
reproductive technology 61
retail 129–30; gravity model 45, 13 *see
  also* shopping
ribbon development 116
Richmond 64
RICS 19, 29, 108, 147, 157
RIG (Radical Institute Group) 166
Rinso Model Village 1
'right type' 158
road safety 177
Rochdale 171, 177
Roman architecture 82–3
Romantic movement 67
Ropemaker Street site ECI 185
RTPI (Royal Town Planning
  Institute) 20; foundation 107;
  headquarters 147; objectives 65;
  Royal Charter 158; 75th Anniversary
  1; teaching unit 111, 180
rubbish tip 134
rural districts 6
rush hour 134
safety 174, 177
Sandstone community 105
sanitary towels 59
Saltaire 93
SAUS (School of Advanced Urban
  Studies) 114

Scandinavia 108, 122
school 42–4; holidays 167; playing
 fields 136
Schuster Report 158
sci-fi 123
scientific: materialism 58; town
 planning 61; -ists 84
Scotland 161, 171
Seaside (town) 168
seats 131, 178, 183
secretary 132, 158;
 clerical 39 see also typists
section 106 agreement 7, 17, 185, see
 also planning gain
segmentation, gender 5
self-employment 130
selfish 54
separate women's super-city 176
servants 90
service sector 127–9, 183
sewers and drains 113; feminism 104 see
 also public conveniences and
 plumbing
Sex Disqualification (Removal) Act,
 108, 112
sexist comments 60
sexual: harassment 50, 60, 155; object
 75; slavery 74
sexuality and space 104–5
Shakers 106
Shakespeare Theatre 122
shanty town 62
Sheffield 170
shire, and rural, counties 26
shopping 45, 79, 103, 129, 132
shopping centres: hierarchy of 183
shops 44, 82, 133, 183 see also retail
Silkingrad 145
'sink for the crockery' 119
sitting areas 131, 178 see also chair
Sizewell Nuclear Plant 64
slavery 74
slum clearance 115, 138, 150
small businesses 130
Smethwick 178
SNAP 150
SOCATECH 169
social: and economic planning 56, 59;
 audit 175; mix 46, 143, 188; policy
 60; services 176; surveyors 145
socio-economic class 39
sociologists 23, 141, 144, chapter 9
Soho 86, 91

soliciting 50
sounds: of children 48, 182; of active
 Europe 46
South Africa 149
South Bank Polytechnic (now South
 Bank University) 175
South Wales Coalfied 126
South West 26, 39
Southampton 16, 171, 177
southwark 176–8; district plan 180
Soviet Union 57–9 see also CIS
Sparkbrook 149
Spinode (cusp) 116
sphinx in the city 72
spirit 8, 51, 71, 105–6
splutter effect 157
sports 181; council 176; jackets 112;
 sexism 50; love of 79; grounds 136,
 139, 148
sprawl 120
square 79
squire's son 136
St Pancras 114
St Paul's Cathedral 129
St Peter's Square 85
steps 43, 47–8, 178 see also ramps
Stevenage 110, 114
Stockholm Plan 109
Stratford-upon-Avon 122
street lighting 48–9, 177, 183
structure plans 23, 140
student loans 100
subsidary wage earners 127
suburbs 43, 47, 142
suffrage 100, 123
sui generis 186
Superman 30
'survey, analysis, plan' approach 144
surveying 19, 29, 129; social surveyors
 145 see also RICS
swathes of meaningless space 48
Swedish flats 118
swords and sorcery 98
Sydney university 163
systems theory 68, 151, 177

Taylorism 119
team sports 137
technician 30, 160
teenagers 43
telamones 83
tele-cottaging 187
tele-shopping 132

temperance 9, 89, 100
Tennessee Valley Authority 144
tenure 41, 180
theosophy 8, 95
thinning out 115
Third World 57, 61, 63, ch. 3
'tilting' 160
time and motion principles (Taylorism) 119
time budgeting 44
tobacco pouch 155
'Tomorrow' 98
Toronto 131
Tottenham 180
tourists 63
tower 123
Town and Country Planning Act 1971 21, 140
Town and Country Planning Association 107
town centre redevelopment 129–31
Town Planning Institute 107, 110 see also RTPI
townscape 65–9
tracers 132
traders 82
traffic: architecture 134; calming 187; lights 134; planning 133
train: commuters 4; disabled passengers 168; spotters 42
transmitters 19, 189
transportation planning 2, 42–4, 53, 133, 135, 169, 174
Tudor Walters 119
typists 4; short-hand 93

UCL (University College London) 112
ultra vires 24, 184
underpass 135
'unfit' 138
'unfulfilled ideas' 148
unitary development plan 6, 28, 184
unprofessional 45
urinals 104
US suburbs 43
Use Classes Order 179
utopias 84, 89, 148
'uze' 22

value-laden fiction 90
vegetables 62, 138

veil 62, 80
verbal pornography 59
verstehen 13
vertical distribution (employment) 25
verzuiling 159
video: Birmingham 170; Sheffield 175; takeaway 46
Vietnam, North 58
village: elders 62; green 136; in the City 150
visionaries 31

waistcoat pocket 83
walk: half a mile 133; '200 yds from bus or car' 131 (see also 177); to work 44
Waltham Forest 177
war-time nurseries 144
war veterans 132
washing: lines 93, 95; machines 59, 64
water: jars 62; taps 62, 114
weeping girls 149
'weed and seed' 115
Welfare State 125
Welsh law 84
West Indies 149
wheel chair access 43 see also mobility; disabled
wheels (buggy) 48
window box 96
wildmen 148
witchcraft 58
Women on the Move 174
wool industries 126
working class 37, 40, 89, 91, 134, 147; community 141; English 89; mothers 150; typical woman 39; 'wrong class' 60
workplace nurseries 180
World Bank 62

ying/yang 72, 75, 78
young mothers 39, 137, 143
Yorkshire 165, 188

zeitgeist 51
zenana 80, 122, 180
zonah 80
zone zappers 57, 61, 63, 70, 81, 109, 139, 162
zoning 13, 40, 81, 127, 117–18